THE LAST WINTER

The Scientists and Adventurers Trying to Save the World

PORTER FOX

First published in 2021 by
LITTLE, BROWN AND COMPANY
a division of HACHETTE BOOK GROUP, INC.

First published in Great Britain in 2021 by
WILDFIRE
an imprint of HEADLINE PUBLISHING GROUP

1

Cataloguing in Publication Data is available from the British Library

Hardback ISBN 978 1 4722 7090 0
Trade paperback ISBN 978 1 4722 7091 7

Offset in 11.5/16.5 pt Sabon by Jouve (UK), Milton Keynes

Printed and bound in Great Britain by Clays Ltd, Elcograf S.p.A.

Headline's policy is to use papers that are natural, renewable and recyclable products and made from wood grown in well-managed forests and other controlled sources. The logging and manufacturing processes are expected to conform to the environmental regulations of the country of origin.

HEADLINE PUBLISHING GROUP
an Hachette UK Company
Carmelite House
50 Victoria Embankment
London EC4Y 0DZ

www.headline.co.uk
www.hachette.co.uk

For Grey, my little fox.

Contents

Introduction

I didn't know these people before I started writing this book. I didn't know about their lives or how they began studying and living in the dark days of winter. I didn't know how their stories added up to a centuries-long worldwide movement to understand ice, snow, and the winter season—and how these elements affect and even control most of the major natural systems on earth. This book is not a cog in that endeavor; it merely documents the effort. So I will not spend this valuable space eulogizing my personal affinity for blizzards, mountains, and the chilling cold that I have spent most of my life in. There will be time for that later. Instead, reader, please turn the page and meet the cast of characters who will carry you through this story: scientists, ranchers, adventurers, vagabonds, time travelers, hunters, and guides. Some of them work at the top of their field; some trudge through the snow just to put food on the table for their families. Some are young, some old. One was alive when I began this book and died on the ice before I finished it. Most live and work in the most inhospitable conditions on the planet. Far too much of their research goes unheralded. These snowscapes are their homes, job sites, and laboratories, and the entirety of their work will determine the fate of human civilization and that of our planet.

Cast of characters, in order of appearance...

Kim Maltais

A third-generation rancher and firefighter living on the eastern slope of Washington's North Cascade Mountains.

Michael "Bird" Shaffer

A self-taught professional skier who has traveled the world with his boards and a parachute and descended some of the most outrageously precipitous mountains on the planet.

Kelly Gleason

A rising star in the earth science world who teaches at Portland State University in Oregon and studies the interface between snow and wildfire, usually on a pair of touring skis.

Jon Riedel

The wizened, soon-to-be-retired official glaciologist of North Cascades National Park—who had the great idea to measure glacial mass balance and movement in national parks back in the 1990s, giving current climatologists baselines and data from which they can extrapolate what glaciers say about our warming planet today.

Seth Campbell

An Arctic ice hustler who has been on sixty polar expeditions in his young career, teaches at the University of Maine's renowned School of Earth and Climate Sciences, and now directs the Juneau Icefield Research Program on the fastest-melting glacier system in the world.

Maynard "M3" Malcolm Miller

A visionary bon vivant who fought in World War II, has degrees from Columbia, Cambridge, and Harvard, was on the first American Mount Everest Expedition in 1963, and essentially invented field glaciology.

Allie Balter

A badass time traveler from Columbia University's Lamont-Doherty Earth Observatory who walks the frozen deserts of Antarctica, Greenland, and Upstate New York, searching for answers about where the last ice on earth will survive.

Brad Markle

Another young gun from Caltech and now the University of Colorado, Boulder, Institute of Arctic and Alpine Research who specializes in paleoclimatology and climate dynamics—reading the tea leaves of miles-deep ice cores to tell us about ancient ocean conditions, climate change, and what direction the wind was blowing on Christmas Day half a million years ago.

Marcello Cominetti

An enlightened Renaissance man and astral soul searcher caught in the body of a ski guide in Italy's Dolomite Mountains—where he once served in the famous Alpini infantry division, then later as a body double in the Sylvester Stallone climbing movie, *Cliffhanger.*

Michael Zemp

The master of ceremonies for earth's glaciers from his headquarters at the University of Zurich's World Glacier Monitoring Service.

Michael Fässler

Former curator of the Swiss Alpine Museum in Bern and keeper of mountain culture in one of the fastest-melting mountain ranges on the planet.

Matt Spenceley

Former British commando and professional climber and ski guide who traveled to Greenland as a teenager and never left.

Rich Manterfield

Benevolent soul and climbing and dogsled guide who ventures into the wilds of eastern Greenland every winter with seal hunters, three dozen Greenland dogs, and a handful of wide-eyed foreigners.

The Hunters

Justus Utuaq, Mugu Utuaq, and Mikael Kunak. Three young Inuit men who grew up on the ice and continue to live off it as their neighbors and nation make the transition into the modern world.

Koni Steffen

The first and final voice on ice melt in Greenland, where he built Swiss Camp in 1990 and alerted the world to the rapid melting there that one day will affect every soul on earth.

The Fires

1

It Started in Cougar Flats

It took fifteen minutes for the pieces that made up Kim Maltais's life to disappear. The stoop where he kicked off his boots after working on the ranch. The faux-wood-paneled hot tub where he soaked on weekend nights with a cold beer. The gas grill on which he cooked lobsters every New Year's Eve, the home where he and his wife, Lenore, had lived for twenty years.

Where the pieces once stood, neatly fitted together on a steep hillside overlooking Washington's Methow Valley, there was only fire and smoke and burning trees. Some said the wildfire that day sounded like a 747 jet engine roaring a hundred feet overhead. Others said it was more like static on an old television, cranked up to full volume. It's hard to say what it was like because the inferno was from another world, something you would see in a movie or read about in the Bible. It was not something that happened to real people, or if it did, they didn't live to tell about it.

If Kim was going to survive, he needed air. Hot wind swirled around what was once his backyard in a dense black whorl of soot and noxious gases. He couldn't see ten feet in front of him. He couldn't hear anything above the roar. He knew that most of his things were already gone; he'd watched the first flame dance up the north wall of his house a few minutes before. The rest would soon vanish. He'd found his cat, Sniper, clinging by all fours on the screen door. The parrot was dead. The fate of his and Lenore's twelve beloved huskies, which had been penned in a chain-link kennel in the yard, was unknown. Kim was pretty sure he'd gotten six out. But the sound of the fire was so loud that he couldn't hear them howling, and they couldn't hear him calling.

Every thirty seconds, another piece vanished: the picnic table, Kim's neatly trimmed lawn, stands of ponderosa pine and lodgepole pine on Pole Pick Mountain, where he had shot gophers for a nickel a hide as a boy. Kim was a son of the West—in stature, diction, dress, facial hair, and every other measure. He was sixty-six and stood "five-foot-twelve" with a chest like a truck radiator. At home he often wore a do-rag, wraparound sunglasses, and a thick gray mustache that, depending on the time of year, morphed into a goatee, pork chops, or a full-on beard. Leaning against a steel fence, he looked like an actor from an old Western—checking on streamflow, fixing irrigation ditches, and surveying his land with a shovel resting on his shoulder.

Even Kim's voice matched the movies: sweet, lyrical, with a northern European inflection infused by his Norwegian ancestors and generations of settlers lured west by the 1850 Oregon Donation Land Act. Who wouldn't sign up for three hundred acres of free land on Frazer Creek? Even in late October, with the bite of winter in the air and the sun so weak you could stare straight into it, the tablelands and hayfields of the Methow Valley were mostly green, and Frazer Creek flowed with crystal-clear water. This was God's country, a perfect sphere of sun, water, gravity, field, beast, and forest.

Kim needed to reconstruct that image—the yard, the driveway, the house, the way out—if he was going survive the firestorm. He'd left an old Jeep in a clearing nearby. There would likely be a couple dozen lungfuls of clean air in it. Hundred-foot flames vaulted through the sky as he felt his way forward. The fire was crowning now, leaping from treetop to treetop. It was so hot that it spontaneously ignited hay, grass, saplings, and fence posts hundreds of feet away.

Fire follows wind more than it does terrain. A wildfire will burn straight down a steep hillside if the breeze is blowing that way. Crown fires run so fast through the treetops, they often don't touch the trunk. Most of the smoke was water vapor, released by the incredibly dense biomass in the Methow forests—up to fifty tons per acre. The rest was mostly carbon dioxide, from torched stands of subalpine fir, Engelmann spruce, ponderosa pine, and Douglas fir. Whatever didn't get hot enough to break molecular bonds and morph into gas drifted hundreds of miles across the Pacific Northwest as ash.

Kim was no stranger to the mechanics of wildfires. Like most young men around Twisp, the little ranch town five miles away, he'd worked on fire crews for two decades. He started with private contractors around Washington, then was hired by the US Forest Service to battle blazes in Oregon, California, Montana, and Alaska. He made squad boss first, then crew boss, then sector boss, managing anywhere from five to twenty crews at a time and staging direct and indirect attacks on fires—back burning, digging fire lines, Kim making the call when to advance and when to get the hell out. Ten years in, he made division boss and directed firefighting efforts across entire regions, marshaling hundreds of firefighters, tankers, excavators, and planes as the blaze cut its path. He'd fought more than a hundred major wildfires before moving home to run the family ranch and live a more peaceful life.

Kim was pretty sure he knew where the fire tearing across Coyote Ridge was headed. It had started days ago in Cougar Flats, north of Twisp. A reporter for the *Methow Valley News* said that she saw smoke on Monday, July 14, 2014. She found Washington Department of Natural Resources (DNR) and National Forest Service vehicles there. Two DNR employees said they were waiting for backup but that simultaneous fires in Stokes, Gold, and Texas Creeks to the south had drained the department's resources. They also said that the Forest Service employee at the scene had inspected the fire and concluded that it would likely put itself out at the top of the ridge.

In just three days, the Cougar Flats Fire was an inferno. And it was racing south. The Okanogan County Sheriff's dispatch received more than a thousand calls a day on Thursday and Friday as the fires spread. Just one day after it started, the Stokes Fire alone grew to six hundred acres, forcing the evacuation of homeowners in Carlton, ten miles south of Twisp. The following day, the fire jumped the Methow River and Route 153 and destroyed eleven homes in Carlton.

The wind continued to blow, and on Thursday, all four fires merged into one massive blaze. What became known as the Carlton Complex Fire grew that day at four acres per second for nine straight hours. The smoke plume was more than forty thousand feet high, and embers from the plume landed up to a mile away, igniting spot fires. By Thursday, the fire had grown by a factor of nearly ten, to 167,000 acres—officially the largest wildfire in Washington history.

Mastodons and Sabertooths

Kim had been relaxing on his boat on nearby Lake Chelan when he got the call that a fire was headed for his home. He spent as much time on Chelan as he could, mostly at a little marina on the southeastern end of the lake. He had never thought of himself as a boat person. He was a rancher who, along with two brothers and a sister, grew up driving tractors, tending crops and cows, and hunting deer so his mother could jar the venison in olive oil for the winter.

Kim's parents arrived in the Methow Valley in the 1930s, just as millions of Dust Bowl farmers abandoned their land and as the nation slid into the Great Depression. There wasn't much to Twisp then. Prospectors had flocked to the valley in the 1880s, when lode gold and silver were discovered. You could pay for your groceries with gold dust back then. But the seams ran dry and the livery barns, restaurants, general stores, saloons, bank, and opera house burned up in a fire accidentally started in 1924 by the town doctor.

Kim attended the local elementary and high schools, played on the basketball team, and learned how to run the farm. Operating a ranch in the Methow Valley in the 1960s was about survival. You fixed what was broken, sold what you could harvest from your land. The Maltaises raised Hereford cattle and cut and sold hay in the summer. If they needed money in the winter, they headed into the forest to log trees.

When Kim was seventeen, he took a long step away from the family business, enlisting in the army and flying halfway around the world to the jungles of Vietnam. If you ask him about his time there, he will tell you that "those are things we are not supposed to talk about." When I asked him how long he was deployed, he answered, "Two years, nine months, six days, forty-seven minutes, and six seconds."

A person looks for solitude after an experience like that, and Central Washington and the Methow Valley are a good place to find it. The same planetary forces that carved Lake Chelan—a fifty-mile-long, 1,500-foot-deep chasm formed by the collision of two glaciers—also shaped the valley. Mastodons and sabertoothed cats of the last Ice Age do not feel far off as you look down the throat of the basin—which starts in the high peaks of the North Cascade Mountains and extends to the prairies hemming the Methow River. The landscape is so jumbled it almost looks alien, perhaps because it once was. Eight hundred million years ago, the northwest coast of

America ended near Spokane, Washington. Everything west of that, including the heaps of gneiss, till, granite, and grassland in the Methow, once formed microcontinents on the Pacific plate. As the Atlantic Ocean grew and the Pacific plate slid beneath North America—it still subducts three inches every year—the islands washed up on the Northwest coast like flotsam on a beach.

You can see the geological turmoil that created the Northwest as you drive toward Twisp. The gnarled peaks of the Sawtooth Mountains near Carlton took shape 300 million years ago, just as plants and reptiles sprouted on earth. The Okanogan Highlands to the north cropped up next, around 180 million years ago, followed by an upwelling of liquid granite that formed the mountains on the eastern edge of the valley. Heaps of terrestrial sandstone, marine shale, and volcanic rock make up the parallel Twisp and Buck Mountain formations framing Highway 20.

Washington Pass is a granite trough; it is the belly of a snake twisting back on itself. It is a time signature on the ever-evolving topographical symphony of the Pacific Northwest. It is carbon, nitrogen, hydrogen, oxygen, and a dozen metals bound in a million ways, cracking earth's crust, eroding under the weight of rain,

From the top of Washington Pass, the head of the Methow wraps sixty degrees to the west, then back around to the east, then west again like a wandering brook. The peaks there are not just tall and steep; they are sheer, like the side of a skyscraper, shooting straight up from the valley floor to forested ridgelines, cirques, bowls, and an auburn rim of granitic summits left by the last glacial retreat of the Ice Age twelve thousand years ago. Down the middle, the Methow River flows, watering alfalfa fields and stands of pine along its shores.

The first time I looked down Washington Pass was a couple of days before I met Kim. The view didn't make sense. Mountains come at you from all directions; steep valleys meet, diverge, then disappear. The stubbled Kangaroo Ridge wanders north-south at a thirty-degree angle to the edge of Liberty Bell Mountain. On the opposite side of the road, the ridgeline between Hinkhouse and

snow, and ice. It is a piece of our history; it wasn't always like this. That is the thing that people, environmentalists, and even preachers forget. They want things to be the same, but things on earth are never the same for long.

Cutthroat Peaks splits and veers toward Cutthroat Pass, then Rainy Pass, where it intersects with the Pacific Crest Trail.

I had not yet unearthed the kernel that this book would wrap itself around. Thinking back on it now, I wonder if that moment and vista was the kernel—a glimpse of the planet and its story, distinct from our human story—and everything I saw next wrapped around *it*. That morning, I'd watched the sun rise above a sea of red taillights on Highway 84 as the lumberjack-hipster apocalypse of Portland, Oregon, faded in the rearview. Straight ahead, the pile of ice and rock that is Mount Hood blocked the sun. It is easy to spot the tallest mountains in the Pacific Northwest. They are all volcanoes and stand on the horizon like mountains a child would draw: perfect triangles with broad flanks and pointy, snow-covered summits. A golden halo wrapped itself around Mount Hood, and an eerie orange glow lit the underside of a marine layer hanging over the Pacific. It was late October—not too cold, not warm, that in-between season when you can wear a T-shirt at midday but your water bottle freezes in the car at night.

I followed the Columbia River for an hour that morning, through a 4,000-foot-deep, 80-mile-long gash in the Cascades that the river cut. You get some perspective on time looking at the Columbia River Gorge and the alluvial flatlands of Central Washington. Some of the greatest floods in the history of North America took place there at the end of the last Ice Age. Massive amounts of water from melting glaciers and glacial lake outbursts on the Clarke Fork River submerged much of eastern Washington with eighty-mile-per-hour deluges. This water eventually hit the Cascades and pounded its way through. Off Route 97, you can still see runnels where the water flowed near the farm towns of Goldendale and Toppenish.

That kind of timescale—geological time, or deep time—is earth's story: its splintered crust and spinning iron core and the abiogenetic miracle that appeared in its oceans millions of years ago, when nonlife turned into life and eventually us. The story of us, even this story that was just coming into view, transpires in regular time—birthdays, graduations, snowstorms, firestorms—which is a blink of an eye in comparison. My guide that first day on Washington Pass, whose arm I grabbed after toeing the comically precipitous thousand-foot cliff that is Highway 20's "scenic overlook," was the Bird Man, or Bird, or Michael Shaffer to Twisp elders who remembered him as a squirrelly blond teenager. A mutual friend had put us in touch, and we made a plan to meet. Bird was forty-nine and a lifetime skier, which meant that he had a lot of free time on his hands May through October—including time to hatch at least half a dozen itineraries for us during my stay.

Michael Shaffer speaks human, but everything else about him is bird. From the moment I emerged from the cell-signal shadow of the North Cascades, text messages to meet for "kawfy" or a "birdfict" bonfire at the local brewpub streamed in. There was a lot to see and many people to meet. The grand finale would be an intricately engineered rendezvous during which I would drive to a designated field, make three loud birdcalls, and wait for Bird to leap off a tiny hill with an equally tiny parachute and land, hopefully, at my feet. Kaw-py?

The instructions looked like this:

1st kaw—I see you, hoot back if you can

2nd kaw—after 5 minutes for setup—you can hoot when ready but hard to hear

3rd kaw—I am taking off!

Bird's parents were also refugees of the Vietnam War, along with many others in the valley, I would soon find. Around the time that Kim Maltais shipped out, Bird's father, Terry, flew F-8 Crusader jets over Vietnam while his mother, Karen, led peace marches around the United States. (The dissimilitude prompted US Navy brass to ask Bird's father to request his better half to "calm that stuff down.")

Love prevailed, and after the war, the couple moved—with Bird, his sister, and seven other families—to the Methow Valley. They bought 240 acres of sagebrush and foothills along Poor Man's Creek and founded the Second Mile Ranch. The Christian-inspired ranch was one of a dozen communes—or cults, take your pick—that moved to the Methow Valley in the 1960s and 1970s. (The name Second Mile Ranch refers to Matthew 5:38–48: "And whoever compels you to go one mile, go with him two.") The families did their best to create a pure and fulfilling life of self-sustainability and community—prominent themes of the back-to-the-land movement that swept through the United States in the early 1970s. Bird and his family lived in a tent at first, then a single-wide trailer home. His father worked as a pilot for a local airline and cut trees to make extra money. The ranch set up a sawmill and, over the next few years, built houses for each family.

Every Sunday, the Second Milers worshipped in church and hosted a potluck meal. Bird had twenty-one surrogate brothers and sisters, by his count. Something about the valley attracted other experimental thinkers. When Bird was young, he and the Second Miler kids climbed the hill behind their houses to spy on the Libby Creek commune—where members hung crystals in trees, walked naked through the fields, and practiced free love without the encumbrance of buildings, relationships, walls, or beds. A few miles in the other direction was the Church of Manna, where the minister and most of the congregation dropped acid before rolling red-hot into a Sunday sermon to see where the spirit took them.

Bird's family did not have much, but the commune members helped one another and made sure everyone was comfortable. Wildfire was a constant reality—many Second Milers worked as part-time firefighters in the summers—as were springtime floods, summer heat waves, and fall droughts. Of all the seasons that Methow residents endured, nothing matched the dark months and deep cold that arrived in November and didn't leave until April.

Winter in northern Washington is something to behold. Temperatures drop below freezing in the high country in September. The first snow typically arrives in October, and the last of it melts in May. The Methow Valley has a high desert climate and sees an average of seventy inches of snow annually. The peaks surrounding it get more than thirty feet as epic winter storms blowing off the Pacific drop up to ten feet at a time.

The cold brings citizens of the Methow together, Bird said, if for no other reason than they chose to live in it and endure it together. While the chill and solitude of wintertide pushes most people in North America south, others move toward it. It is not a masochistic endeavor. It is an appreciation for nature's rhythms: the wobble of Earth's orbit, the passing of seasons, the moment the planet's axis tilts back 23.5 degrees, creating the shortest day of the year and the beginning of winter. Butterflies, bees, and eighteen hundred bird species head for the equator then. A few hundred varieties of fish, insects, and cetaceans follow. Caribou walk a thousand miles over five months. Arctic terns fly the equivalent of twice around the world in their migration. Overhead, trees redirect chlorophyll, which gives leaves their green hue, to their root systems, allowing other pigments like xanthophylls (yellow) and anthocyanins (red) to show through—until the leaves fall, completing the transfer of energy from sun to the soil.

Winter is not a weather event. It is, in part, the result of an ancient astronomical collision. A planetesimal (tiny planet) struck Earth soon after the larger planet formed, four and a half billion years ago, knocking Earth off its 90-degree orientation to the sun and creating the cadence of our seasons. The Moon is a piece of Earth's crust that was ejected after impact; the drag of its gravitational pull slowed

Earth's spin. Before the crash, daytime lasted six hours longer, and winds, caused by the planet's spinning beneath its atmosphere, often reached speeds of three hundred miles per hour.

The flick of Helios's finger, which pushes half the planet into darkness and cold for half the year, also ejected most of Earth's primordial atmosphere into space, much of which was carbon dioxide. As a result, our planet does not have the extreme carbon dioxide concentration and resulting greenhouse effect of our sister planet Venus, where surface temperatures average eight hundred degrees Fahrenheit. Earth's uniquely thin firmament lets solar energy come and go and allows water to exist in all three phases—liquid, solid, and gas—necessary for life.

Before we knew about planets, orbits, and celestial obliquities, the Greeks hypothesized that cold and winter came from a distant source, beyond the realm of Boreas, god of the north wind. Aristotle wrote that an area of frozen ocean somewhere west of Europe was the culprit; medieval priests said it came from the westernmost island in the known world, called Thule. Scandinavian epics describe the extreme cold of the north as the home of giants, death, drowning, and "strange beasts." Some of the last Neanderthals to live in Ireland built a mass tomb around 3200 BC to honor the first day of winter. On the winter solstice, the sun shines through a roof box in the tomb at the end of a sixty-two-foot tunnel. Beneath the opening are stone basins filled with cremated bodies.

For most of history, the first days of winter marked the beginning of famine, dark days, freezing temperatures, and no crops. Feasts and celebrations on the winter solstice were thrown not in anticipation of this miserable season but because it was the end of the harvest season, when crops were reaped and wine and beer were fully fermented. Larders were full; death was at the door; livestock was slaughtered so animals wouldn't have to be fed throughout the winter, leaving an abundant supply of meat. It was a time of preparation, when days began to grow again, and a new year began—marked by festivals like Maghi in India, the Norse-Gaelic-inspired Hogmanay in Scotland, and Yaldā in the Middle East. Saturnalia in ancient Rome was one of the most extravagant solstice celebrations. A sacrifice was made to Saturn, gambling was temporarily allowed, and masters switched roles with their slaves, serving them food and drink at a public banquet where gag gifts were exchanged.

In the ancient Greek doctrine of the five zones—demarcated by the Arctic Circle, Tropic of Cancer, the Equator, Tropic of Capricorn, and the Antarctic Circle—winter existed only in "Frozen Zones" set in the far north and south. The "Torrid Zone" awaited wayward travelers on either side of the equator. If someone wandered too far north, the person would freeze to death immediately. Those who ventured south would burst into flames. "Temperate Zones" in between were the only regions humans could survive. The theory of the five zones was so pervasive that it limited exploration and trade for millennia—until sailors broke through the borders looking for new lands and trading partners. Prince Henry the Navigator of Portugal sailed into the Torrid Zone off the African coast in the 1400s and continued a bit farther south each subsequent year. The prince not only proved that the Flaming Zone could indeed be crossed but also showed that there was some fine land, resources, and even people down that way. The Irish headed in the opposite direction, discovering Iceland in AD 795. They found no "region of universal death," either. Instead, they found people living on the forested island quite happily. Icelanders, in turn, explored the west coast of Greenland, the icy mothership of the North Atlantic, up to Smith Sound, a couple hundred miles north of the Arctic Circle. A thousand years later, Europeans chased whales and seals to Spitsbergen and the Arctic Circle, and the idea of winter and varying climates was cemented.

Seeing winter overtake the North Cascades is like watching a set change onstage. The scene couldn't be more different from summer—when grass grows and dogs bark and screen doors slam shut. Rules that warm-weather folks take for granted are reversed in winter. Streetlights shine up through airborne ice crystals instead of down. Gravity softens in deep snow, and sounds are muffled several decibels as billions of crystals thicken the air and cover pastures, highways, and hilltops. When Bird was young, he and his friends dragged sleds up the steep hillsides surrounding the ranch, then rocketed down to the flats. They built snow forts and snowmen and shoveled paths for their families. Eventually, Bird ditched the sleds for a pair of old Nordic skis, then alpine skis that a Lithuanian neighbor had given him.

Residents of Twisp didn't have to travel far to find a ski area. Since 1958, volunteers had cut, cleared, built, and managed a small ski resort on 5,375-foot Little Buck Mountain, fifteen miles outside town. This is what winter people do in winter: invent ways to pass the time. It was not easy; it took thousands of unpaid hours to keep the resort running. The community appointed board members to manage the hill, held fundraisers to pay for the resort, sold tickets to keep it open. Local architects drew plans for the base lodge, and builders, plumbers, and electricians constructed it.

By the time Bird started skiing at Loup Loup, there was one rope tow for beginners and a Poma lift that dragged everyone else twelve hundred vertical feet to the top. Skiers had their pick of ten runs to come back down. With no farming, logging, fishing, tourism, or really anything else to do in the winter, people traveled from all over the Methow Valley to meet up and ski the Loup. A ski school cropped up, then a ski racing program. Season-pass holders were mostly from town and could just as easily be found elbow deep in the machinery of the Poma lift as they could schussing down a groomer. A ski patrol shack and a maintenance building eventually appeared, along with all the trappings of a little American ski-resort-that-could. In 1998, the board added the Loup's pièce de résistance: a used four-person chairlift that a nearby ski area had discarded. Over the next year, for better or worse, volunteers installed it themselves.

Again, Loup Loup is not like any other ski area you have ever seen. There is no snowmaking, no high-speed lifts, no midmountain wine bars. The only on-mountain amenities are a few toboggans stashed in the woods to haul injured skiers to the ski patrol shack and the Little Buck Café in the lodge, where you can get biscuits and gravy for six-fifty or a Polish sausage for five bucks. To anyone envisioning Aspen, Colorado, or Zermatt, Switzerland, or any other alpine resort where highborn tourists in turtlenecks sip hot toddies between runs, erase that image from your mind. Replace it with a picture of Bird swinging his skis on the chairlift while Kim Maltais bombs through the glades below on a pair of Kneissl White Stars, clumps of snow clinging to his beard and a Pabst Blue Ribbon jammed into his jacket pocket. Because that's how it was and how it still is at the Loup. It is also where our stories about Bird and Kim and wildfires collide, right here at the snow line, where fall turns to winter, water turns to ice, forests grow and burn, and the alchemy of the Frozen Zone extends its reach into the civilized world.

Kim's family lived on the road to the Loup, so all he had to do to ski was walk to the end of his driveway and stick out his thumb. His first skis came from the trunk of a 1955 Pontiac Bonneville that his father bought from a Swiss family living in town. The seller was a skyline logger and kept a pair of skis, poles, and double lace-up leather boots in his car. The skis were solid ash with cable bindings and were twice as tall as Kim. He learned to ski on them and how to screw the steel edges back on at night. For most of the 1960s, Kim was the first on the hill in the morning and the last off. After the closing bell, he cleaned the lodge with the manager, Roly Brinkin, and caught a ride home in the bed of Brinkin's 1953 Ford pickup, along with half a dozen bags of trash they had hauled from the lodge.

Kim earned his first pair of modern skis—Toni Sailer "Fibreglaskis"—by working at the resort. (He currently holds the oldest season pass in the valley: 1965.) Loup Loup was open only weekends and Wednesdays. If Kim was missing from school on a winter Wednesday, his teachers had a pretty good idea where to find him. He skipped school so many times that the principal started waiting at the lift line to catch him. Kim's basketball coach didn't like his winter hobby, either, and told him he had to pick between the two sports. "That was my last day as a Twisp Yellowjacket," he said.

Kim's father joined him on the hill on volunteer days. He fixed the Poma lift and lent logging equipment to mend and install lift poles, pull cables, and clear trails. In a county with a population density of eight people per square mile, Loup Loup was the de facto winter social scene. Showing up to volunteer was like showing up for the Fourth of July parade: if you were around, you didn't miss it. "Work parties" involved everything from reshingling the lodge to painting lift poles to thinning glades and digging a fire line around the resort.

When Kim returned to Twisp after the war, Loup Loup was his winter sanctuary. He skied all season long, split wood for the fires, and chipped in wherever he could. One year he drove his camper to the back of the parking lot and set up a winter shanty he called Sheep Camp. The camp is running strong today. The primary mission of its members is to keep the fire pit going and to empty as many beer cans in between ski runs as possible. It was on one of those icy runs, in 1981, when Kim schussed past a cute skier, hockey-stopped next to her, and asked the young woman's name. "Lenore," she'd said. And yes, she would love to ski a run with him.

On this peaceful autumn day at Loup Loup, I found a disaster. Dozens of people sat hunched over on the ground, holding their wounds. An entire family lay on their backs under a pine tree, moaning. One or two ski patrollers knelt beside each one, triaging, taking pulses, and performing CPR. It was terrifying—until I saw instructors and clipboards and realized that it was a ski patrol training session.

"Goddamn Climate Change"

It was Lenore who called when Kim was sitting on his boat on Lake Chelan. It took him five minutes to close the hatches and about forty to make it home, driving twenty miles an hour over the speed limit. After seeing a DNR crew posted up across the street, he went straight to his parents' ranch house to check on Lenore. Kim's parents had passed away years before, and the house doubled as a storage unit for family heirlooms. He set up the irrigation system to wet the fields around the home as well as the neighbor's house across the street.

He had asked the DNR crew to dig a fire line and use their pump trucks to keep the blaze from jumping Highway 20, but when he drove back to his and Lenore's home, the crew was sitting on their tailgates, taking selfies, and talking on their phones. The fire had not only jumped the highway by then but also jumped the creek. The presence of the crew was a false assurance, leaving him and his house exposed. A half hour later, Kim found the Jeep, climbed in, and waited for a clearing in the smoke.

He made a few false starts and retreated each time. The heat made the Jeep's steel frame tick, and smoke filtered into the car. He looked for the huskies, listened for a howl. Fire had encircled the property, and the air was pitch-black. Flaming limbs and trees fell to the ground as the blaze ripped through his home, pouring out the windows and doorway. Family photos and keepsakes burned, as did his clothes, guns, computer, tools, skis, and ski boots. In fifteen minutes, it would all be gone—a lifetime of memories in the only home he and Lenore had ever lived in.

A sliver of light opened up, and Kim put the Jeep in gear one last time and lurched downhill. The kennel was directly to his right. There was no time to stop now. To the left, flames rolled up a twenty-foot drop-off. A dozen trees stood between him and the driveway, but somehow he missed them all and found his way to the dirt road. He followed the bridge over Frazer Creek and made his way onto the highway. The DNR crew sat there, watching. The fire was now headed for his parents' ranch house. He pulled into the driveway and fired up an excavator to plow a trench between the creek and the highway. Back and forth he went all afternoon, raking up grass and cutting a fire line around the home.

The blaze was a giant orange cloud roaming east, then west, a phantasmic beast following a weather system it had created. Just before reaching the hay barn, the flames skipped back over Frazer Creek and ran straight up the hillside away from the ranch house.

Fifteen minutes later, it turned again and headed back down. By the time it jumped the creek a second time, it was east of the house.

A single patch of green grass surrounded Kim's family home. Everything around it was charred or on fire. There was no time to feel relief. Kim knew the area could reignite at any second. He continued hosing down the property, digging fire lines, and helping the neighbors do the same for three straight days. His only clothes were

The thermometer in Kim's backyard. "This was all cooked, all on fire," Kim said. "And I had no idea where the dogs were. I couldn't hear 'em; they couldn't hear me. They were panicked... One ran in the hills for about two weeks. People said they saw it, and they'd heard it, but it wouldn't come to anybody. I came back down here one day, and I walked in the pen, and I looked. There weren't even any doghouses left, but there was this brown blob. I thought, 'Good Lord, what could this be now?' You never know. I called the dog's name, and it moved and turned around and came out. It came back home. Its paws were burned, and it was hungry, of course, but she's still alive, still got her."

the ones he had been wearing on his boat. The fire had burned everything else, including the power lines to town. The old ranch house had no water hookup, no working toilets, and nowhere to sleep. Boxes and piles of furniture filled every room. Lenore and Kim found a mattress late that first night, dragged it into the garage, and fell asleep.

More than three hundred families slept in someone else's house that night. Five hundred homes and structures burned in the fire, including half the nearby town of Pateros and one in ten homes in Twisp. Damages amounted to $30 million. (Half of the three hundred homes that burned were reportedly uninsured.) Kim's brother lost his house, as did several neighbors. The house next door that Kim had helped with was saved. In all, the Carlton Complex Fire burned 256,000 acres—an area five times the size of Seattle.

If you have seen a forest fire or driven through the charred moonscape that it leaves behind, you know it is a not a singular event but a geographical shift—on the scale of an earthquake or a volcanic eruption. Nothing is the same after, and the blackened deadwood and torched earth remain for decades. The thing about the Carlton Complex Fire in particular, and the puzzle that I was still piecing together, was that it was not the last fire, or even the last natural disaster, to hit the Methow Valley. Just three weeks after Kim started cleaning up his ranch, a thousand-year flood dumped an inch of rain in an hour and washed out the entire Frazer Creek basin—including Kim's driveway, farm shed, and large sections of the highway. The following summer, fire hit again when the Okanogan Complex inferno torched three hundred thousand acres, forcing the evacuation of Twisp and Winthrop and killing three young firefighters a few miles from Kim's house.

A similar trend played out across other mountain ranges and towns in the West. Fires ravaged Breckenridge, Colorado; Sun Valley, Idaho; South Lake Tahoe, California; and Bend, Oregon. In

2017, a million acres burned in Montana, and the following year, 1.4 million acres burned in British Columbia, Canada. One study pointed out that 60 percent of all fires in the American West since 1950 have ignited in the last twenty years.

The size of the fires was even more concerning. Since the 1980s, the average magnitude of wildfires in the United States has doubled. In Western states alone, more than 2 million acres burned in large fires between 2002 and 2012. The cost of fire suppression more than doubled as well. Between 2006 and 2016, damages spiked to $5.1 billion. According to Kim, the media tried to connect the rise in wildfires and extreme weather in the West to "goddamn climate change." But seventeen of the eighteen warmest years on record—all of them occurring since 2001—coincided with a sudden increase in fires. Temperatures in the US West were rising at several times the global average. Even more troubling, the surge in fires may have been tied to the one thing that kept Kim sane all these winters. Since the 1970s, the rate of *winter* warming in the West had tripled, replacing snow with rain, and reducing Western snowpacks by 20 to 50 percent—drying out boreal forests in the process. In the West, 50 to 70 percent of all precipitation is stored as snow, which as it melts releases water throughout the spring, summer, and fall. These days, warmer air temperatures melt snow weeks or months earlier in most ranges, creating drought conditions and increased spring growth that adds fuel to the fires. The eight most fire-ravaged years in recorded history had all seen historically low snowpacks.

As with many frightening scenarios in the realm of climate change, the situation appeared to be spiraling out of control. The length of winter is projected to decline across the United States, in some locations by more than 50 percent by 2050 and by 80 percent by 2090. Spring snowpack depths across the country are forecast to drop during the same period by 25 to 100 percent, according to the National Oceanic and Atmospheric Administration (NOAA). In the big picture, it makes sense. Put an ice cube in a warm room, and it melts. But the process of global warming is more complicated. A sudden melt would kick off a long line of natural disasters and internal feedbacks, including a greater spike in forest fires, which in turn would release more carbon into the atmosphere and increase warming and melting—possibly beyond the point of no return. Which is to say that winter, snow, and ice are more than sources of seasonal fun—they are fundamental components of the planet's heat and water cycles that we might not be able to live without.

These revelations were hard to comprehend. How could that much snow disappear in forty years? How could the frozen summits of the Rockies, where I had skied for four decades, transform into brown, muddy heaps in the winter? And the Alps? The Andes and Himalayas? The poles? It didn't seem possible that someday it could all be gone. There was too much ice on the planet—twenty-seven quadrillion tons of it on Antarctica alone.

Climate predictions and study plots were far from Kim's mind in the final hours of the fire. The end of his world had already come and gone. Six dogs were missing, including the one that wandered into his yard a month later. The view of Pole Pick Mountain from his parents' home was blackened forever. The blaze was now ripping up Little Buck Mountain. All the fires had converged into a twenty-square-mile circle of smoke and flame that moved at speeds of up to sixty miles an hour. The dot in the center of that circle was Loup Loup, which, Kim realized, was likely in ashes. In fact, as hot wind and embers arced over the summit and the grass under a weather station there caught fire, someone or something flipped a switch. The wind changed direction and blew the fire back downhill, saving the Loup in one of the few silver linings of the day.

2

The Bird Is Sick

Driving into Twisp does not feel like driving into a ski town. Visitors are met with a junkyard, several mobile homes, and a cluster of log cabins set along the fast-moving Methow River. Les Schwab Tire Center, across the street from eleven bright yellow school buses of the Methow Valley School District, occupies the next block—followed by the Methow Valley Masonic Lodge and a bright red cinderblock lean-to that is Twisp Feed & Rental, known for its crimson marquee one-liners: "My broom broke, so I drive a stick."

A skinny beanpole of a kid walking around the NAPA Auto Parts lot next door, wearing knee-high muck boots and a waterproof camouflage raincoat, summed up the vibe of the town: quiet, rustic, somewhat insular. Looking at the line of television antennas and radar dishes above Second Avenue, all pointed at the pale blue freezing sky, I got the feeling that certain locals could live without the intermittent stream of nature lovers, like me, causing backups at the town's only traffic light.

Most tourists don't come to Twisp for the skiing. They come for the Western feel and unspoiled countryside. What locals call foothills would be entire mountain ranges back East. Driving south past the kitschy Western facades, emporiums, and old-time photo parlors of nearby Winthrop—or "Win-throw-up," as Kim calls it—you half expect to see John Wayne hobbling down the street with a six-shooter on his hip.

Indeed, the actual Wild West is not far off in time or distance. The Oregon Trail, which delivered hundreds of thousands of migrants to the West in the nineteenth century, is just two hundred miles south. Before that, the Methow tribe lived in the valley for ten thousand years, sleeping in woven-mat shelters, paddling cedar dugout canoes, and fishing the Methow River with spears and weirs. The tribe suffered from the same diseases and deception that decimated most other American Indian nations. They were protected by the Columbia Reservation in the mid-1800s until President Grover Cleveland abrogated the treaty in 1886 and opened the Methow Valley to mining and settlement by whites.

At least one beaming face, at the Twisp River Suites, was happy to see a visitor drive into town. The suites were, as far as I could tell, the only hotel in town aside from a row of cottages on the north side of the river. The suites are set in a brutalist, patinaed steel structure that looked like it could have been a recycling center or maybe a server farm. The hotel was an afterthought, I soon learned. Most of the units were built as condos, until the owner realized that the type of person who wants to live in deep winter for eight months of the year is not the same kind of person who wants to be sandwiched in an apartment building with a shared balcony. When nobody bought the condos, the hotel was born.

Ms. Trinity welcomed me at the front desk, which was set in what apparently had very recently been someone's living room. She had been waiting for me for days. There was so much to talk about, so many bits of information that would make my stay not only enjoyable but also safe and efficient. She jumped right in with a brief history of the town and the hotel, followed by the layout of the town's primary road systems and public services, the overall operation of the hotel and its various amenities, a detailed weather report for the next five to fourteen days, and some suggestions for what to do with my time when I was not enjoying my suite. This tidal wave of particulars was just a primer, I soon realized, a bit of foreplay before the main event: the world-famous breakfast bar that occupied the western wall of the lobby. The list of what would and would not be available at the buffet, and when—Seasons change! Tastes change!—was extensive: dried blueberries, pine nuts, yogurt, eggs, bagels, a homemade lox schmear with the smoked salmon and capers already mixed in.

Trinity's tour continued on a balcony strung with Christmas lights, now layered with frost. At the far end, tantalizingly close to the room I had booked, were a barbecue and hot tub. The grill and tub were, in fact, owned by the owner of the hotel and were not to be used by anyone but him, unless permission was granted, which it was for me, thankfully, because I happened to be staying in the owner's personal condo while he was away.

I had, at this point, been holding my bags for so long that both of my thumbs had inexplicably gone numb. What I wanted more than anything was to *see* the inside of my room and to *be* inside the room. My wish was granted as Trinity mercifully opened the sliding glass door to a charming suite with a propane fireplace, couch, kitchen, and bedroom. Everything I needed, and then some, was there, including a second bathroom in the hallway, Trinity said, in case I felt like mixing things up.

The workings of the room were explained in an extremely high level of detail, the last of which Trinity offered as she backed out of the suite step by step. I could still hear her advice—on how to operate the coffeemaker, the dishwasher, the sink faucet—after I closed the door, delicately, in her face. The warmth and serenity of the suite was not oversold, and I spent the next two hours on the couch, alternately staring at the North Cascades through the window and pondering what I was doing here. I had come to document winter in North America and investigate the connection between climate change, disappearing snow, and wildfire. But I had more questions now than when I began. Would climate change snuff out winter completely? If winter disappeared, what else would we lose? What would a place like Twisp look like in a century? What would the West look like? What about the Arctic? Or the culture and tradition of winter that I had grown up with? Finally, just how long did we have before winter was gone?

Bird saved me from this notional cloudburst with a text message: Are you Kawming? The bonfire was on, chili was simmering, and Twisp's winter royalty was gathering right then at the Old Schoolhouse Brewery Taproom to talk about life in the North Cascades. I had completely forgotten about the meeting and scrambled to get dressed, hop into the car, and drive to a building filled with shiny school-bus-sized vats of hops and beer. I almost didn't recognize Bird when I arrived. He was perched on a wooden fencepost, cloaked in an Obi Wan–like hooded robe that nearly hid his face and small frame. It didn't seem like the kind of garment you were supposed to wear in public. He said it came from a thrift store in Japan. His ski buddy Kobayashi had found it. His prescription wraparound glasses and a hooded sweatshirt pulled tight beneath the robe's hood made for a look so disarming, so frighteningly *bird*-like, that I didn't know what to say and blurted out, "Wow!"

A few bearded men shuffled around a bonfire in a gravel beer garden. One was Owen, a 220-pound brute with soft eyes and a furry grin. He had just come from a Loup Loup work party where he'd thinned out glades on a kid's trail so the kids could learn to ski in the trees. Next to him was Jason, who had hair erupting from nearly every pore on his face. Jason had been a bad-boy mogul skier in the 1980s, then disappeared to Aspen Highlands every winter thereafter to work and ski. The third wise man, the one I ended up talking to—or listening to, I should say—was Ed the Carpenter. He had the bony face, slender calloused hands, and 1960s puffer vest—which he'd picked up at a secondhand store just hours before—of a man who made a living with his hands. He was Jesus-like in appearance, and it struck me that if Jesus had made it to retirement age and happened to wander into Central Washington, I wouldn't be surprised if he followed Ed's lead, bought a snowmobile and a pair of skis, and spent sixty days a year skiing bottomless powder.

There were skiers in Jesus's time and even well before. The oldest skis on record date to 6300 BC, excavated by Grigoriy Burov in the Vychegda River basin of Russia, 750 miles northwest of Moscow. The skis were preserved in a peat bog near Lake Sindor on the site of an ancient Mesolithic civilization. On the bottom of the ski is a carved moose head, which Burov suggested could have been used for traction going uphill or as a symbol of speed going downhill. There are also mentions of skiing in early Chinese, Roman, and Scandinavian writings. Rock carvings like the 4000 BC "Rødøy Hunter," discovered in 1929 in Norway, depict camouflaged hunters on long boards holding a single pole, likely racing downhill toward large game mired in the snow. The etchings and early descriptions suggest that a certain mysticism surrounded the sport and refer to hunters, warriors, kings, and even gods cascading down a mountainside, staff in hand, ready to inflict a lethal blow on enemies or prey. The similarity in ski design in geographically remote places is remarkable, like concavities in the ski base or small pieces of wood that held the foot in place—in Norway, Latvia, Slovenia, China, Siberia, Korea, and even California. (Sierra goldminers were obsessed with ski racing in the 1850s, becoming the fastest-moving humans on earth as they hit speeds of eighty miles an hour on quarter-mile tracks.)

Most people who arrived at Bird's bonfire that night lived

completely within the folds of the skiing life, having left their homes, jobs, and, sometimes, loved ones to deep-dive into winter living. Their access point was the North Cascades Highway, which closes to automobiles November through May and offers a fifty-foot-wide trail straight into the mountains, which possess some of the greatest backcountry skiing in North America. If you could see a time-lapse of the transformation from fall to winter there, it would look something like this: western larch fading to gold in a green sea of subalpine fir, pinegrass and elk sedge turning brown, vine maples radiating red as white alder and huckleberry pale to mint green. Sometime around mid-October, the first of a hundred winter storms to hit the contiguous United States every year spins off the Gulf of Alaska, spills down over the North Pacific, and charges above the rain forests and drainages of the western Cascades into a wall of rock at Washington Pass. As the warm, saturated air rises, trillions of droplets in the clouds freeze around airborne dust. More droplets bond to the frozen nucleus, forming six sides, six branches, and, eventually, a snowflake—the icon of winter that graces Christmas wrapping paper, pom-pom hats, carved oriels in Bavarian chalets, and everything else that has anything to do with the myths, traditions, industry, and even demons of the winter season.

A peculiar quality of water allows a snowflake to form. Water molecules are asymmetrical, with two hydrogen atoms on one side of the oxygen atom, giving it stronger polarity that aids in forming hexagonal crystals. It can take up to three days for a snow crystal to fall to the ground. A cloud bank can hold more than a million gallons of water, so if conditions are right, it doesn't take long for snow to stack up—micrometer-size droplets swirling in the cloud, a supernova of creation spitting out the most multifarious and short-lived crystals on the planet. Storms that come in over warm air—fifteen degrees Fahrenheit is the sweet spot—produce the most snow. If it's too warm, you might see thunder snow, a meteorologi-

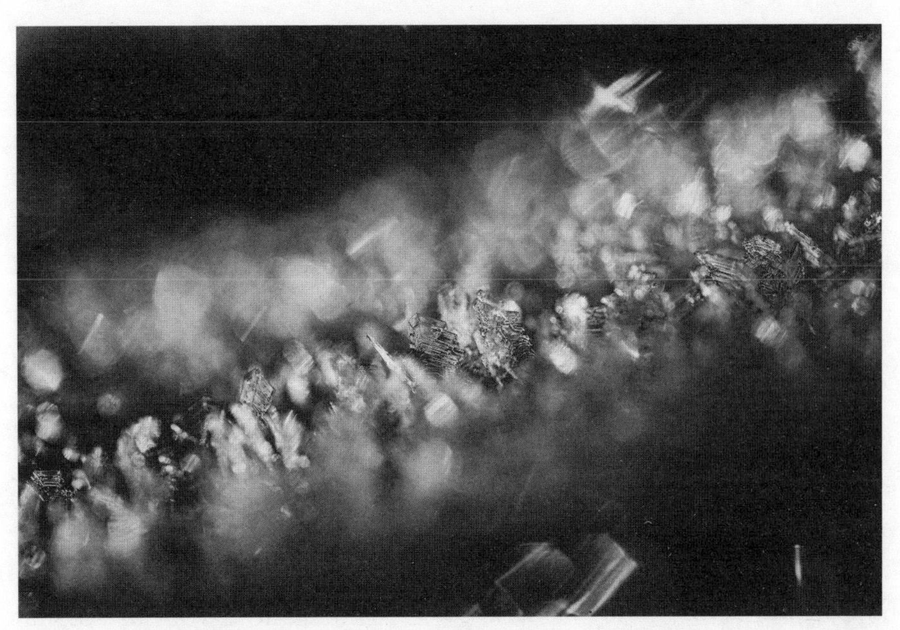

Dave Heath, a longtime senior Powder *photographer, took this picture and many others of snow and the skiing life in North America. I hiked with Dave, Mike Hattrup, and Mark Newcomb into the Taurus Mountains of Turkey on an assignment. We skinned past troglodyte caves for a few days and eventually skied some surprisingly steep and long couloirs. After we hiked out, we spent a week driving dirt bikes between ancient cities on the Mediterranean for fun.* (Dave Heath)

cal acid trip of an event where turbulence creates thunder and lightning in the middle of a blizzard. Snow can fall with ground temperatures up to forty degrees, as evaporation of melting ice on the airborne flake refreezes it on its way down. The largest snowflake ever recorded was two inches across, though it was likely a polycrystal composed of several flakes stuck together. Some of the smallest forms of snow are called graupel, which exists somewhere between hail and snow.

After snow lands, it can take many shapes. Sastrugi form when wind gouges a snowpack, carving lines and divots. Penitentes grow up to ten feet tall and were first described by Charles Darwin as he was crossing a mountain pass in the Andes. Snowy megadunes are to Antarctica what sand dunes are to the Sahara. But it is powder, defined as an unspoiled layer of fresh crystals not yet deformed or melted, that draws skiers in.

What Bird, Kim, Ed, Owen, Jason, and most other skiers in Twisp can do on a powder day is hard to define, because it breaks the rules that most of us grew up with. The slopes that Bird skis after a winter storm—gracefully, like a real bird dipping through the sky—are so steep that they become waterfalls and vertical cliff faces in the summer. With twenty feet of powder caked to them, they are a semisolid cushion that you can slide through or leap forty feet into without consequence. Which is what Bird does all winter long, often alone, gliding down through the crystals, a single track appearing behind him like a curl of smoke.

Growing up, Bird skied every day he could. He was a natural, one of the best skiers at the Loup. Years later, he became one of the best skiers in the world—navigating death-defying slopes in the Alps, appearing in ski magazines and Warren Miller ski movies and, yes, garnering his name for his natural predilection for flight. A visual aid is needed here to illustrate the level of human–winter communion we are talking about, so I point you to an online video featuring Bird in the steeps of the French Alps. The video opens with a haunting electronic ambient soundtrack and shots of the Aiguille du Midi—a spire of rock and ice that protrudes two vertical miles above the French town of Chamonix. The Aiguille is the stage on which the greatest skiers in the world perform, and Bird describes the line that he will ski that day: the Frendo, a spine of snow that starts on a precipitous razor-thin ridge and comes to a sudden end at a thousand-foot cliff.

Because the tram on the Aiguille was broken, Bird drives to the Italian side of the mountains and hikes up. (This whole episode takes place between breakfast and lunch.) It is here that he informs the viewer that the Aiguille is "Mama" and Bird is ostensibly her

Look closely. Closer. The dot in the middle of the upper snowfield is Bird. (The dot on the left is his pal, Sami Modenius.) There is a plume of snow behind Bird and a ski track winding up to a traverse. The face is Col du Plan, near the North Face of the Aiguille du Midi. I have skied some hairy lines in my life, some near this, but this one is ridiculous. I actually get a fluttering sensation in my chest looking at it. Feel it? (Damien Deschamps)

son. A not-so-subtle Freudian theme emerges as Bird walks on touring skis toward the iconic, remarkably nipple-like summit of the Aiguille.* (The French have long associated mountains and breasts.

* Touring skis release at the heel and use climbing skins for traction—to walk uphill in the backcountry. At the top, skins are removed and the heel is locked down to descend.

See Wyoming's "Grands Tétons.") Mama probably wouldn't have approved of the line that Bird skis off the north side, disappearing into a series of steeps, each more hideous than the last. There is no trail. Only a few dozen people have ever skied the line, which is more commonly known as a technical climbing route for alpinists headed up. A human body accelerates on an icy slope at pretty much the rate of free fall, which the aptly named law of falling bodies describes as thirty-two feet per second per second. Halfway down the face, it indeed takes Bird about a second to straight-line a vertical thirty-foot river of black ice. He is going so fast at the end that it takes a hundred yards to stop. Snow swirls and a GoPro mounted on his helmet nearly bounces off as the hazy green pastures of Chamonix Valley rapidly approach. Then Bird rests for a moment and calmly unfolds a small parachute from his pack, chirping, "Let's fly the flock out of here!" He then skis off the cliff and glides leisurely back to town. The film closes with him back in Chamonix, arms back, chest out in a *Titanic*-esque swan pose, followed by a very Bird-like sign-off: "Chamonix, France!"

Things weren't always this beatific in the skiing world. Bird had lost friends to avalanches, crevasses, and falls. It was part of the lifestyle, part of skiing on the edge. The odds are not in your favor in the mountains, especially in winter. Years ago, Bird recalled, he was on his way to ski a classic route called the Gervasutti Couloir in Chamonix when two young skiers from Bozeman, Montana, started down from the opposite side. The chute is essentially a tongue of ice hanging from the summit, wider at the top than it is at the bottom. Bird was planning to ski the classic entrance from the left side. The other group was skiing the more challenging right-hand entrance.

Bird spotted a streak of black ice in the middle of the line before one of the Montana skiers slid onto it and began to slip. The man shouted back to his partner that he couldn't hold. Bird watched, speechless, unable to help, from a hundred feet away. Then the man's ski tips skittered out from under him, and he tumbled thousands of feet to his death. "We didn't ski it after he fell," Bird said. "I couldn't stop thinking about that for a while."

The Blizzard of Aahs

I was familiar with elements of Bird's world. My family had also moved north in the 1970s. We left Upstate New York in 1975 to settle in northern Maine. The Köppen-Geiger climate classification system defined our home in New York as "humid continental," which means a few months of real winter and about seven months of camping weather. Köppen-Geiger classified our new home on Mount Desert Island as "warm summer," which loosely translates as "the only time it is warm is eight weeks in summer." A local confirmed this description a few days after we arrived. "You got two seasons," he said. "July and winter."

Muddy snowbanks lined Route 3 the day we drove in. I don't remember this—I was three years old—but was told about it by my mother, who, a terrific skier and winter adventurer herself, periodically regretted the move for the first ten years. All winter, nor'easters blew the front doors of our house inward and blocked the driveway with heaps of snow. New parkas were purchased; L.L.Bean boots were fitted; cordwood for the stoves was split. Week by week, we learned how to exist in our new, winterized world, and in the heart of winter, when it got dark at three in the afternoon, we learned to ski.

The sport ran in my family: my grandfather had been an avid skier and taught his five children how to turn on a thin layer of hay in their backyard in Philadelphia. I hated skiing with a remarkable passion at first, until I turned twelve and felt the confluence of vectors that make up a proper ski turn. Like Kim and Bird, I became obsessed and worked various jobs to afford new skis and lift tickets. I cut out exercise routines from the back pages of ski magazines,

My mom (with my dad) ran off to Austria in 1966 when she was nineteen to be a ski bum in St. Anton. She found a room in a family-run Bavarian pension where, every morning, the older couple who owned it served her an egg, salami, cheese, and bread. She enrolled in Ski School Arlberg, which Austrian ski sensation Hannes Schneider founded in 1921, and explored the slopes of St. Anton, Lech, and Zürs. Europe was booming after rebuilding from the war—just two years before the Paris student uprisings and the Prague Spring. My mom spent most afternoons lounging on the decks of the Hotel Post, Krazy Kanguruh Bar, or the Bahnhof, where food and wine were cheap and young people from all over the world mingled.

then performed the bizarre hopping and squatting sequences in my room. In high school, my parents rented a run-down, one-room cabin a few miles from Sugarloaf Mountain for $500 a season. There was no running water. Heat came sporadically from a kerosene furnace salvaged from World War II. My mother, despite her love of skiing, was understandably so opposed to the state of the cabin—and using the outhouse in the winter—that my parents stopped coming. The cabin was unceremoniously handed down to me and a group of teenage skiing hooligans who became my best friends. We practiced our craft during the day, skiing bumps so tall that we vanished into the troughs between them. Daffies were in, mule kicks out. I broke a finger, dislocated both shoulders, and sustained my first concussion learning to ski fast and throw tricks. This was real winter, with windchills of minus eighty-five degrees Fahrenheit on the mountain regularly.* At night, Budweisers vanished by the hundred as we stayed up as late as possible—in part to avoid freezing to death when the heater inevitably died. The loft in the shack was so hot it was hard to breathe, and the ground floor was so cold you would wake up with frost in your eyelashes. The only cooking device was a toaster oven. Stouffer's pizza, Pop-Tart heaven.

Our inspiration was Greg Stump's ski movies, specifically *The Blizzard of Aahhh's* and the smooth skiing of Scot Schmidt, Glen Plake, and Mike Hattrup. We bought the clothes they wore, and we tried to emulate their airplane turns, steep skiing moves, and cliff jumps. I wrote my college essay about climbing and skiing the East Gulley on New Hampshire's Mount Washington and attended one of the only schools in the country that maintains its own ski area. I signed up for ski patrol my first semester and bought an eighty-nine-dollar season pass to Mad River Glen. I worked on the patrol through Christmas break to make extra money and slept in a dorm not far from the farmhouse where Robert Frost wrote his dark, enigmatic verse about snow. I kept a calendar in my dorm all four years to keep track of ski days. One January, I logged twenty-eight days.

---

* My good friend and mentor, ski photographer Wade McKoy, traveled back to Sugarloaf with me once to shoot a feature there. He had just tested his cameras at a deep-freeze facility in New York for an upcoming trip to Antarctica. They passed the test but froze on the top of Sugarloaf.

Four friends and I moved to Jackson Hole, Wyoming, after we graduated. The resort received forty-two feet of snow our first winter. Minus-twenty-degree days were not uncommon. We stacked our skis in the trunk after every storm and drove up Teton Village Road to get in line for Jackson's tram. The mountains were massive and steep compared with what we knew, and we had to learn to ski all over again. There were few buildings along Highway 22 then, just cattle range, the slate-blue Snake River, and the sudden upthrust of the Grand Teton Mountains cutting a north-south line along the Wyoming-Idaho border.

While friends joined investment firms in New York City and started law school, I skied a hundred days a year and put on a Domino's Pizza delivery uniform four nights a week.* (The job came with a discount ski pass.) The following year, I got another job as a cub reporter for the *Jackson Hole News*. My beat included girls' junior varsity volleyball games, school board meetings, and, occasionally, skiing. I sent a few stories to *Powder* magazine, the bible of the sport that I'd been a subscriber to since I was thirteen years old. I rarely heard back. I had long fantasized about traveling overseas with a pair of skis and a reporter's notebook. Every fall, I looked for daubs of white paint on a relief globe in the Teton County Library, indicating snowy peaks in exotic mountain ranges: India, Chile, Russia, China, Papua New Guinea. I sent dozens of ideas to *Powder,* and years later, perhaps out of fatigue at reading my submissions, an editor called. He liked my pitch about skiing the Indian

* Some images from that time: bottomless, untracked aprons of powder falling away for a vertical mile; launching off a forty-foot cliff, landing and skiing away; skinning through granite-lined chasms deep in the backcountry; skiing through trees at thirty miles an hour; pool balls flying through the Stagecoach Bar and Grill; reject pizzas stacked five-high on an oven; clear, starry skies above a winter tent site; standing on the roof of a Jeep going fifty miles an hour; skiing with six friends at midnight through a spruce forest with a handle of Jack Daniel's tucked into my jacket.

Himalayas and wanted to know if I could do the trip for $2,000. I had never seen two grand in the same place in my life, and I immediately answered yes! The following week, I packed everything I owned into the back of my truck, moved out of the room I'd been renting, and, a month later, flew to New Delhi with my skis. An image I will never forget: waking in the middle of the red-eye and spying the lights of Tehran sparkling in the pitch-black desert below.

For the next two months, I carried my skis on trains, rickshaws, and antique DC-10s with ski photographer Wade McKoy and a crew of two snowboarders and another skier. McKoy had shot for *Powder* since the 1970s and taught me about ski mountaineering, reporting, and traveling. He carried manual, steel-bodied Nikons and grew out his hair and beard on the trip—completing the "on assignment" fantasy I had conjured over the last five years. It took a few days to drive to the Garhwal Himalayas, thirty miles north of Rishikesh. There we hired twelve porters to help us climb and ski the 16,000-foot headwaters of the holy Yamunotri River.

Pilgrims had hammered a thousand coins into a sacred tree near base camp, offerings to the Goddess Yamuna, the snow, and the meltwater that sustained millions downstream. We waited out a three-day storm that dumped four feet of snow on our tents. When the skies finally cleared, we summited the pass and walked from the frozen world we'd been trekking through for a week into lush green vegetation growing on the southern side. It was there that I recognized for the first time the effect of snow, meltwater, and winter on our lives. A gust of hot wind hit my face. The brownish plume of New Delhi loomed on the horizon. Rivulets of water gathered and joined on the hillside, eventually forming the second-largest tributary of the Ganges River. After skiing back to camp, we smoked hash with the porters—which turned out to be opium—and taught them how to ski, all of us screaming with hilarity.

I filed the Himalayan story that spring, and *Powder* offered me a job in the fall. The job description was to spend three months of the winter finding the most exotic snowfields in the world. I bought a world relief map for my office and pinned more white peaks: Bolivia, Japan, Russia, New Zealand, Italy, France. I discovered a massive, unknown ski resort in post-communist Bulgaria where you occasionally had to bribe a lift operator to fire up a chairlift. I skied a sliver of snow on a ten-thousand-foot peak in Baja, Mexico, and vanished into neck-deep powder in Hokkaido, Japan—where ten feet of snow fell during the ten days we were there.

One year I flew to Istanbul with my childhood idol from *Blizzard of Aahhh's*, Mike Hattrup, for a two-week expedition into Turkey's Taurus Mountains. Killing time on a four-hour layover, we sat on a carpet in Istanbul's Blue Mosque, ski boots around our necks, hundreds of worshipers kneeling on rugs around us. The following season, I saw Bolivia's Chacaltaya, the highest ski resort in the world, then skied first descents from 15,000 feet in the Cordillera Quimsa Cruz range. A couple of years later, we climbed and skied 19,600-foot Toqlaraju in Peru, where I spoke with a farmer about how his village had to move because water from the receding glacier no longer sustained it.

A young skier and journalist, Hans Saari, introduced himself to me during the Peru trip. He was writing about another peak in the range, Huascarán, far taller and vastly more difficult than what we planned to climb. Saari was a good writer and had sent me a few stories to consider over the years. After we returned home, I accepted his pitch about skiing classic ski-mountaineering descents in the Alps. He set out that winter with a team of young American ski mountaineers. The trip got off to an inauspicious start in the French Alps with bad weather and dangerous snow conditions. When the weather finally cleared, he and photographer Kris Erickson headed up from Chamonix to ski the Gervasutti Couloir. They were excellent ski mountaineers—prepared and capable. Saari headed into the couloir first, where he encountered a rib of black ice. It was there that Bird watched him slip and fall to his death—and where the filaments of Bird's life and my life intertwined.

It would take twenty years for Bird and me to meet—and two hours into our second interview to realize our shared loss. It took the arrival of a *new* life, my daughter Grey, and a decision to take on this odyssey after a long break from skiing to bring us together. In the decades in between, Bird continued his rise through the pro skiing world, knocking off terrifying descents and adding a parachute to his quiver. I drifted in the opposite direction, away from the mountains and skiing. I moved to New York City, attended graduate school, fell in with a rowdy crowd of artists, traveled the world on homemade boats, got married, and traveled the world again with my wife, Sara. I still wrote about skiing and even reported on climate change and its effect on snow, but the thrill of the sport had faded, and my days as a skier-journalist were over.

My days as a father had just begun. I walked to my newborn daughter's crib every morning before work and held her as she rubbed her eyes and nuzzled my neck. I held Grey's hand as she learned to slide down the stairs on her butt, then shoot down the slide at the playground. I sat in my office during the day, transcribing interviews for assignments and paying bills. The web that Sara and I had woven to keep our family safe and comfortable was so complex, so tightly knitted, that if one thread fell out of place, everything fell. These anxieties disappeared in the dim red glow of the night-light, as I looked down at the fuzzy lump in the crib—too big, too much hair, practically a toddler now.

One night after reading an article about the formation of our planet, I stood over Grey's crib and envisioned streams of atoms, neutrinos, and electrons flowing through her. She was being assembled, an inch at a time. She was made of stardust, as was I. I remembered seeing her in the delivery room a few minutes after she was born. It was like the overview effect that astronauts describe when they look at the fragile bubble of Earth floating in space. I did not know the little shape under the heat lamp. She was still blue; she had yet to take her first full breath. Her eyes were wide open, and she was staring straight at me. I was oddly embarrassed, not sure how to introduce myself. I should have rehearsed something, memorized a poem or a quote to commemorate the moment. But I just stood there, glancing back at Sara, waiting for someone to tell me what to do. Grey wiggled her hand and breathed in. Her body flushed pink as oxygen filled her lungs, and she reached toward me and grabbed my finger. She had a surprisingly strong grip. I remember thinking, I must protect this being.

Grey was born with a million eggs inside her—ready to pass life on to another generation. If she is a late bloomer like me, she will bear her first child around 2060. Earth will likely have warmed three degrees Celsius by then—if humans don't stop burning fossil fuels. Sea levels will have risen by more than one and a half feet, flooding coastal cities from New York City to Shanghai. Most of the Amazon rain forest will be dead, along with the world's coral reefs. Half of the earth's population will experience three weeks of lethal heat every year, displacing a billion people from equatorial regions, radically increasing freshwater scarcity, and cutting crop production by a fifth. With billions of migrants forced to look for a new home, mass migration and resulting armed conflict could very well consume Asia, Europe, and the Americas.

I didn't want to know about the potential environmental devastation; I would rather have remained naive. But after reading and reporting on it, I could not unsee it. One day, as I walked home from the physical therapist's office, I caught a glimpse of how it would affect my life. Forty years of skiing had worn a hole in the cartilage of my right knee, and Grace Gronkowski was rebuilding it after two surgeries. By complete chance, she had also been a skier for most of

This is what a forest looks like for thirty or forty years after a burn. People move on, and the burn just sits for a full generation before the forest finally starts to "green up," as Kim Maltais put it.

her seventy years. Her husband was a coach on the US Ski Team. Dr. Gronkowski played with Grey while I pedaled an exercise bike and hopped around on one leg. The first snow of the year fell that November day as I pushed Grey back to the car. Tiny crystals collected on her sleeve, and she reached out to grab one. Toothless grin. Flakes in her eyelashes. Time was running out. The Alps were melting, the West was burning, ice sheets were vanishing. Watching Grey at that moment, I knew I couldn't stop the end of winter. But I needed to know when and how it would all play out, and if she would get to experience it the way I had.

3

The Transfer of Energy

High on the flanks of Mount Adams, above the deep-time floodplains of south Central Washington, an experiment was underway. It involved blaze-orange poles, pile drivers, circuit boards, climbing ropes, closed-cell foam, tiny metal shields, weather instruments, and a plastic box of nearly every kind of screw, hex bolt, washer, nut, and fastener ever collected. The site was a campground before the hillside burned to the ground. The US Forest Service somehow still considered the Morrison Creek Campground operational and maintained signs pointing to latrines, parking areas, and all the backcountry bureaucratic bric-a-brac that makes the Forest Service run—such as the little wooden box where campers deposit reservation slips and another where they can get trail maps of the area.

But there were no campers here. There seemed to be no life at all, and by the look of the charred black landscape, there wouldn't be for decades. The forest that once divided and sheltered the campsites was no more. The only trees still standing were either bleached white or charcoal black. They stood at odd angles that live trees do not ordinarily assume: fallen halfway to the ground, snapped in half, reduced to jagged and broken trunks, or wrenched into a semicircle as if a surge of gravity had pulled the top down toward the soil.

Around the campsites, rangers had cut other half-burned trees, which now circled the area with a tangle of deadwood. One blessing the disaster afforded was an unobstructed view of Mount Adams—itself a cauterized heap of igneous rock that reached two miles into the sky. The same subduction that created the mishmash of geology in western Washington also opened a few thousand volcanic vents along the Cascade Volcanic Arc, from which magma occasionally flowed to twenty major volcanoes. The last time Adams blew was around a thousand years ago. Estimates for its next eruption are unknown, though volcanologists said it would likely be a *flank sector collapse,* similar to the 1980 Mount St. Helens eruption.

The rocky, eighteen-mile-long base of the stratovolcano rises from the forest to scree-covered ridgelines and alpine bowls. From the flat summit, you can see other volcanoes along the arc: Mount St. Helens, thirty-seven miles to the west, and Mount Rainier, fifty miles to the north. During the last Ice Age, which began around 115,000 years ago, glaciers completely covered Mount Adams. Twelve glaciers on the mountain survive today, including the White Salmon Glacier, which we could see arcing down the southwestern face in a river of gymnasium-sized seracs.

Kelly Gleason wasn't fazed by the geological war zone where our group had assembled. It was eight in the morning, and she was more concerned with the health of another ancient being: her Subaru, which annoyingly refused to die, thereby preventing her from getting a new car. The auburn wagon was parked in the dirt lot, noticeably low to the ground and very much broken-in, with saggy seats and a collection of maps, other papers, coffee mugs, and her daughters' toys crowding the dash. "I park it on the corner every day, hoping someone will sideswipe it and I can get the insurance money," she said. "But no one ever does!"

The life of a scientist is largely about patience—waiting for results, grants, breakthroughs, anything to further our understanding of this world—so Dr. Gleason was used to biding her time. She wore the unofficial uniform of a snow scientist: black microfiber long underwear, wicking zip top, soft-shell trekking pants, and technical hiking boots. Her long brown hair was held back in a braid beneath a black trucker's cap.

Gleason was a lifetime skier, and on a backcountry ski run near the Three Sisters region in Oregon's Cascades, she made the discovery that changed her life. The lack of tree limbs and the abundance of light in burn areas makes for deep snow and great skiing. Gleason noticed something else about the snowpack, though. Its consistency seemed different. At the bottom, she scooped up a sample and noticed black specks in it. The tiny bits of carbon from burned trees could absorb sunlight better than white snow could. This black material, she realized, was a conduit for the sun's energy. She inspected the snow near chunks of charred deadfall and burned trunks, which, along with the ash, were cumulatively melting the snow faster than normal. Multiply this effect across millions of acres of burned forests around the West, and, Gleason speculated, a major feedback loop was possible. If more forests were burning, and those burned trees made the snowpack melt faster, then these forests would further dry out and burn more easily and melt more snow. This self-propagating cycle could affect major alpine forests across the West, the ecosystems downstream of them, and even the atmosphere—as burning forests emitted carbon instead of absorbing it.

Gleason was then working for Anne Nolin, a snow science pioneer who has worked with the National Aeronautics and Space Administration (NASA), among others, since 1997, innovating and performing remote sensing analysis on frozen terrain around the globe and on planet Mars. Dr. Gleason and Dr. Nolin were studying the effects of climate change on watersheds in the Cascades, and Nolin was mentoring the up-and-coming scientist on everything from remote sensing techniques to grant-writing. Gleason's path to snow science had not been a straight or easy one. She was gifted in elementary school, skipped grades, and aced science and math courses. Physics and trigonometry were her favorites. She barely passed ceramics. As a kid in Seattle, she spent most of her free time outdoors. Her family didn't take vacations; they hiked to alpine lakes and skied at Snoqualmie and—one of my favorite ski hills—Alpental. After her parents divorced, Gleason moved around. At one point she ended up with her mother in Arlington, Washington, closer to the North Cascades and Mount Baker.

Divorce can scramble a kid's compass, and Gleason found herself mixed up with a rebellious crew of older snowboarders in high school, skipping school most winter days to ski at Alpental. She moved back in with her father in Seattle when she was sixteen and ditched classes again to drive her old VW Bug—orange chassis, baby blue fender, cobalt fender, purple hood—to the mountains to ski. She stopped studying and let her grades slip. "If I hadn't had that healthy, self-empowering place, I would have been a very different person," she said. "I may have barely graduated high school because I spent way too much time skiing, but it saved my life. I mean, it made me a whole person, and I made much wiser decisions because I had mountains in my brain."

This isn't a book about Gleason's secrets. (Her words.) It is a book about how winter, snow, and ice shaped our world and even us—and how its coming end will change everything again, in every season. It is also about how people like Gleason are trying to figure out how and when all this might happen so that the rest of us can prepare for—or even slow down—climate change. Suffice it to say that after entering the scientific world and taking on dozens of years-long projects like analyzing "the vertical distribution of epiphytes in a nine-hundred-fifty-year-old hemlock forest on the east slope of Mount Rainier," Gleason ended up in the Himalayas, monitoring prescribed burns by Tibetan yak herders. It was there, in a hazy cloud of smoke, grazing pasture, and glacial runoff, that she decided to study the relationship among snow, water, and fire.

A wildfire is a transfer of energy—from the sun to trees, then trees to the atmosphere and earth—ultimately conveying carbon, hydrogen, and water. In the early 2000s, a scientist by the name of LeRoy Westerling pioneered the study of how snow affects forest fires. Dr. Westerling's research found that the occurrence of large forest fires in the US West had increased by 500 percent since the 1980s and that a vast majority of those fires burned in areas that were once covered with spring snowpack. Fires in seasonal snow zones were four and a half times larger than ones outside the zone and boosted the average burn area in the West by over 1,000 percent. Where an average fire in the 1970s burned for six days, fires between 2003 and 2012 burned an average of two months. Fire in former snow zones influenced fire season length as well—doubling the season since the 1970s. For every increase of one degree Celsius in average global temperature, firefighters, farmers, families, skiers, and everyone else living in headwater regions of the Rocky Mountains could expect a 3 to 700 percent rise in burn area, largely because of melting snowpacks.

These numbers are just averages. In hot spots, like the once-snow-covered forests of the Northern Rockies, the situation is far worse. Wildfires there are 3,000 percent larger than they were in the 1970s. As snowpacks thinned in the Pacific Northwest, including the forests of the Methow Valley, fire frequency increased *1,000 percent,* and burn area grew by *5,000 percent.*

It doesn't take a degree in hydrology to understand the mechanics of how a lack of snow, and thus water, can dry out a forest and make it more susceptible to fire. But as with most events in the world of climate change, the problem compounds. Lack of snow cover allows sunlight to sustain new plant life earlier in the spring—adding more demand on the shrinking water supply and more fuel for fires. With warmer temperatures increasing evaporation from the forest floor—which was once covered with reflective snow but now is dark, heat-absorbing soil—the domino effect is turning massive swaths of the West into a tinderbox.

Just watch the nightly news to see burning towns, mountains, and cities across the West and around the world. In the final months of 2019, wildfires burned a thousand homes and twelve million acres in the United States. The next year was worse, with a record eight million acres burned in the US West and more than ten thousand structures razed. Fires in California torched nearly four and a half million acres, twice as much as the previous record, including five of the six largest fires in the state's history. The hottest decade on record was 2010 to 2019, followed by the decade before that and the decade before that. In 2019, historic fires also ignited unlikely places like Alaska, Siberia, and western Greenland, pointing to another alarming trend: the Frozen Zone was warming faster than the rest of the world.

To get a picture of just how much snow the planet has lost in the last century, we have to pan back, way back, into space and observe the white polygons of Earth's cryosphere—defined as places on Earth's surface where water exists in solid form. White patches of snow and ice grow from the poles as winter freezes midlatitude regions. Since the mid-1800s, though, when humans began burning massive amounts of fossil fuels, the edges of each polygon have been deteriorating. A million square miles of spring snow cover has vanished from the Northern Hemisphere—*since the 1970s*. Between 1972 and 2015 in North America, snow-covered terrain has diminished by thirty-three hundred square miles per year, and high-elevation snowpacks in the western United States have decreased by 41 percent, or an area the size of South Carolina, between 1982 and 2016.*

* Ice is melting too—for example, on the Great Lakes, where annual maximum ice cover is 22 percent less now than it was fifty years ago.

You would think climate change would melt snow evenly, the way your freezer thaws when you leave the door open, but melt varies widely from region to region. A warmer world will, in fact, see increased precipitation—the atmosphere can hold about 4 percent more moisture per degree of global warming—bringing *more* snow to some areas in the short term. Many climatologists believe that a warmer world will also disrupt the jet stream, antithetically creating "polar vortex" events like the 2021 deep freeze that iced over highways in Louisiana, dumped snow on the Mexican border, and plunged Texas into a subzero blackout—all while the planet heated up. In the long term, the thawing freezer analogy works better—winter shrinks and precipitation falls as rain instead of snow.

Either way, climate change is devastating winter. Rocky Mountain snowpacks are down by an average of 20 percent. Spring snowpacks in the Cascades have diminished by 20 to 40 percent since the early 1990s. By the end of the twenty-first century, climatologists predict a further 65 percent decline in spring snowpack in the Cascades and California's Sierra Nevada—if emissions continue on their current track. According to a 2020 study by the European Space Agency's Climate Change Initiative, North America lost forty-six metric gigatons of snow per decade for the last forty years. (One metric gigaton of ice would fill Central Park in New York City to a height of eleven hundred feet.)*

Finding further ramifications and feedbacks of melting snow in and around America's forests was Gleason's newest task. She marked out a study plot in a mixed conifer forest in Oregon's High Cascades that had burned in the 2011 Shadow Lake Fire. She measured snow-

* Remember that "their current track" is not completely predetermined; it is a choice. The door to a cooler world is still open. In the last ten years, the once-prohibitive cost of solar energy, gridwide battery storage, electric cars, and wind power has plummeted, making the switch from fossil fuel to renewable energy entirely possible.

water equivalent (how much liquid water exists in a snowpack), solar radiation (how much sunlight makes it through the trees to the snow's surface), reflectivity (how much of that light is reflected back into space), and snowpack surface debris (like soot, which can absorb more sunlight). She found that even though snow accumulation was greater in the burned forest, burned debris made the snowpack melt twice as fast as an unburned forest. Another study, in 2019, put a finer point on the results: burned forests in the West were absorbing around 350 percent more of the sun's heat for up to fifteen years after a fire. Combined with the increase in fires in seasonal snow zones, the West was seeing a fourfold rise in the amount of sunlight absorbed by mountain snowpacks and drier summer conditions. The melt-drought-burn-melt-drought-burn cycle was indeed self-propagating.

Gleason grabbed a water bottle from her Subaru and let me in on another danger awaiting our crew. We were in cougar country, she said. She seemed excited about this. She once lived in a remote cabin in the mountains with her husband, a writer, who was working on a book. A family of cougars living beneath the cabin was raising a litter. Gleason and her husband also had a cub, their daughter, who was a toddler at the time. Early in the morning, the human family would watch the cougar family creep across the yard, slaughter a deer, then drag it under the house and eat it for breakfast.

The story contained more notes of horror than it did charm, and Gleason seemed amused by my frightened reaction. She was driven by genius, curiosity, and tenacity, all of which gave her a unique way of relating to actual people. Death by cougar was, in fact, how she wanted to say farewell to this life, she said. "Imagine the last moments!" she howled. "What a way to go!"

Gleason was the most experienced scientist at the campground that day, but she was not in charge. That responsibility fell to a young woman named Katie Swensen. As a PhD student at Washington State

University, Vancouver, Swensen served as a hireling to her adviser, a distinguished WSU scientist named Kevan Moffett. Dr. Moffett studies "heterogeneous, nonlinear, and dynamic coupling of physical and biological flow and transport pathways in the hydrological cycle," or how water moves through the atmosphere, earth, and plant life. She had been studying the interaction between snowmelt and forest growth and needed a series of time-lapse cameras and monitors to be installed at a backcountry study plot.

Swensen was about as excited to perform her duty as a carpenter is to demo a bathroom—which is to say, not unexcited but, rather, compelled by duty more than passion. She wore a black hoodie with "Buffalo Geology" on the front and skiing bibs with pockets on the chest for pens and scientific tools. A somewhat deadened stare suggested that, although she might rather be working on an experiment of her own design, nothing was going to stop her from trudging five miles out and back to complete the job. She was a mountain woman, comfortable in mountain terrain, including the growing windstorm right then overtaking Mount Adams.

The third member of the party seemed oblivious to the lenticular cloud taking shape over Mount Adams. Jesse Krause was perhaps the most interesting man in the Pacific Northwest. While most members of the snow-science world fit the mold of alpine skier or outdoor enthusiast of some sort—or, at the nerdiest end of the spectrum, amateur birder—Krause was none of these. His five o'clock shadow was pushing eight or nine o'clock, and his glasses were so thick that it was hard to see his eyes through them. Beyond these superficial details, Krause's defining characteristic was a fountain of knowledge so vast—including biology, astronomy, opinion, and utterly random bits of observation and allusion—that the group treated him more like a computer than a human, teasing out calculations and relationships between disparate events. There were times that day on the long and surprisingly strenuous hike when I would have rather stabbed out my eyes with a pen than listen to Krause say another word. And yet every time he spoke about glacial till and ponderosa seedlings or explained how local Native Americans timed the beginning of winter with the moment the first alpine larch trees dropped their needles, the more I revered him.

We dutifully followed Swensen to a picnic table in the parking lot, where she had attached all the necessary poles, ropes, foam, and instruments to four metal-frame backpacks, which meant that I would be carrying one. "Put something blaze orange on it," she said. "It's hunting season." She then led us to the South Climb trailhead and east through a bramble of charred hundred-foot tree trunks. There was no actual trail. Above our heads, thick ribs of snow, ice, and volcanic rock fanned down from Mount Adams's summit. The weather forecast called for extreme wind all day, which could make walking through a charred forest extremely dangerous. A third of the burned trees on the slope had already fallen. "Keep your eyes up," Swensen said, alluding to the fact that eventually all of the trees would fall—and you did not want to find yourself between them and the ground when they did. In a lengthy email a few weeks earlier, Swensen's adviser had clearly spelled out the dangers of walking to the study plot. There would be forty-five-degree slopes, downed tree trunks, brush to fight through. The windstorm had already doubled in strength by then, accentuating Dr. Moffett's final warning: "Please put as your first priority to look out for each other, and for weak, burned trees that might be prone to fall." Already, I had witnessed twenty-foot branches, treetops, and massive hunks of bark crashing to the ground.

Krause and Swensen filled me in on the project as we walked. The time-lapse cameras, snow-depth measuring poles, and various weather sensors would record the depth of the snowpack as it rose and fell throughout the winter. Soil-moisture monitors and time-lapse images of surrounding plant life would show how melting snow watered the forest. Parts of the site we would be studying were in a *triple burn* on the southwestern slope of Adams—a swath of forest that had burned three times in a row. It was a unique study plot, Gleason said, given that a triple burn there would typically take two to three hundred years. Here it had happened in fifteen.

The World Before This One

The windstorm grew all morning. Thankfully, the sun was bright and warm. Water trickled through a streambed, and the greenery of an unburned swath of trees appeared even greener against the blackened forest around it. "This is how it is supposed to work," Gleason said, pointing to the live trees. Snow in the winter, rain in the spring, the last of the snowmelt watering the forest through the fall. Every now and then a lightning strike hit the right place at the right time and a section of the forest burned. Flames triggered the release of heat-sensitive seeds that germinated, took in water, sun, and minerals from the ashes, and the whole system started all over again.

It had been a perfect cycle for much of the planet's history—born of its orbit, composition, tilt, and atmosphere. Around 90 percent of the solid structure of a tree is formed from invisible gases in the air. Energy from the sun splits water into hydrogen and oxygen inside a seedling and combines this oxygen and hydrogen with carbon from carbon dioxide to form sugar. This process, what we call photosynthesis, serves as the foundation of plant growth as the sugar molecules combine to form cellulose and lignin. Trees then develop root systems that work in symbiosis with local fungi. Energy absorbed into the tree is released into streams and the soil by falling foliage and deadfall. Shade from limbs and leaves allows for cooler land and water. The cooler, moister land also promotes life and, eventually, an ecosystem that forms around the base of the tree.

The deadfall that we were trying to navigate bordered on impassable. My newly rehabilitated knee felt as if a bag of quarters were sliding beneath my kneecap with each ascending step. The surgeon had made a serious mistake during my first follow-up exam, casually showing me an iPhone video he had shot of the procedure. The clip—think chainsaw massacre meets high-end deli counter—is now burned so deeply in my psyche that every click and pop of my knee reminds me of the New York strip steak that is my right quadriceps.

Gleason commiserated. She had torn her anterior cruciate ligament two years before, also skiing. Her doctor gave her the same dismal prognosis that mine had given me. After surviving a year of surgery, extreme agony, and gimp-legged recuperation, she would have to continue physical therapy, and would be in pain, for the rest of her life. My doctor's suggestion for the enduring pain: "We could do it over?"

I hobbled into Swensen's first snow-study plot forty minutes later. It was marked by fifteen-foot blaze-orange poles jutting out of the ground. A seven-foot PVC pipe striped with black and orange electrical tape stood above a spiderweb of blue conduit, which disappeared in the earth to measure moisture in the soil. A weather station to read precipitation, humidity, and other data was mounted on the pole and was connected to a data logger box, a circuit board, a battery, and a small solar panel that powered the system.

The apparatus's data, analyzed alongside weather data, allowed Moffett and her team to see water arrive as precipitation, get stored in the snowpack, be released into the soil, and finally get taken up by plant life. Our job—Krause's privilege, I should say, because the man seemed to relish the assignment—was to attach temperature and humidity modules to trees surrounding the study plot. The tiny instruments were protected by what looked like tinfoil birthday hats, cut and taped together by Swensen at home, and were supposed to deflect wind, sun, and rain for a more accurate reading.

The contraptions were fragile and awkward, and somehow Krause made it even more awkward, white-knuckling a wobbly aluminum ladder and screwing them to trees. He wore a plaid lumberjack's jacket—a real one, with a sweat-stained collar and frayed cuffs. He had chosen treadless work boots for the job, perfect for slipping off the rungs when a gust of wind shook the tree. The entire situation made it almost impossible for him to lean into the power drill and secure the hats with drywall screws—leaving Gleason and me scrambling around the trunk, trying to hold the ladder in place, occasionally handling Krause's backside to keep his mass from plummeting down on top of us, and helping him leverage his weight to make the drill do its work.

The wind did not die, nor did Krause, thankfully, so the job went on, hour after hour, tree after tree. Studies like this are about the most tedious, intensely focused tasks imaginable, like stacking a thousand paperclips, over and over, for months, years, decades at a time. And doing it without getting lazy or bored or forgetting the big picture, as it were, of what the data you are harvesting will ultimately mean to your study and possibly the human race. The overriding assignment: watch things change over long periods to glimpse into the deep-time crystal ball.

This is what Aristotle and Leonardo da Vinci recognized when they noticed that patterns of fossilized mollusks in the mountains were strikingly similar to living ones on the seashore. It would take an extremely long time for the latter to become the former, which meant, as the eleventh-century Chinese naturalist Shen Kuo wrote in his deep-time treatise, that another world existed before ours and that many more existed before that. Without this kind of understanding, we are looking at a children's story of our planet: swaying trees, pointy mountains, sparkling glaciers. With it, we begin to understand how things work—as well as the massive sea change coming our way.

The operation underway at the foot of Mount Adams was a tiny cog in this process of understanding. Swensen, Gleason, and Krause seemed uninterested in the significance of the puzzle; they were more titillated by process and were engaged in a silent and intricate forest dance—methodically hooking up cameras, checking CPUs, screwing in space hats, and taking soil samples. I had nothing to do but watch, and my mind wandered to the millions of tons of deadwood dancing about above us. I took shelter for about forty minutes behind a giant ponderosa stump, hoping it would protect me. A half mile below, at the edge the obsidian ghost forest, was another living forest. It cut a green line across the hillside, then angled west toward a streambed. Slats of yellow sun cut through the clouds, lighting the ravine and feeding the trees. Twenty miles away, above the canopy, I could see a wide, blue bend in the Columbia River. There were no houses or roads in sight—just a loamy, dark green ecosystem working in perfect synchronicity. It had survived the timber rush at the turn of the twentieth century, when Washington State turned out five billion board feet of lumber every year. It survived the hydropower and irrigation booms, when engineers constructed more than four hundred dams throughout the Columbia River basin. What was happening now was different, though, and it was affecting, quite literally, everything.

A seventy-foot pine creaked in the wind as the gale ratcheted up to what felt like the beginnings of a hurricane. The sound of wind through the dead branches became a shriek, then a roar. The summit cloud above Mount Adams was now a microfront. The temperature dropped fifteen degrees, and tiny raindrops peppered the southern flank of the mountain. The trunk holding up the tree in question was 95 percent burned through. "It's been like that for two years," Swensen said without looking up.*

* Swensen emailed me that after a winter of unbelievable storms, it fell a few months later!

Jesse Krause (right) knows more about this planet than you do. Kelly Gleason (left) is a close second.

Perhaps to distract me from impending doom, Swensen allowed me to pound an iron rod into the ground and fit it with the last blaze-orange measuring stick. Then, unceremoniously, she said that we were done, and we all raced down the mountain. Krause was a welcome distraction for forty minutes as he explained the journey of a water molecule down Mount Adams—from a snowflake to the White Salmon Glacier to the White Salmon River, then the Columbia River, through the Columbia Gorge, and finally past Portland and Astoria and the forested beaches of what Meriwether Lewis and William Clark contemptuously named Cape Disappointment into the Pacific. Along the way, the river supports coniferous wetlands, white oak forests, salmon, trout, and hundreds of species like beavers, pond turtles, woodpeckers, spotted frogs, and, of course, humans.

Krause was still talking about the river of life when I branched off to follow Gleason on what looked like a safer route. It was, and a few minutes later, I saw the glint of a windshield. Krause had wan-

dered into a dead end and was battling his way through a web of fallen trees. Gleason, Swensen, and I scurried into the welcome opening around the parking lot a few minutes later and took our packs off as the wind toppled a thirty-foot tree on the edge of the burn. I thought about the old adage of a tree falling in the woods—and the hubris of humankind to consider that, if no one saw it, it didn't happen. Every tree that falls, burns, or dies affects everything in the world. In the realm of deep time, we are barely breaching the warmer epoch that climatologists now call the Anthropocene, or the Age of Man. Another danger for trees in this new world, Gleason said, took place recently in a forest in New Mexico. It burned several times and after the last blaze, when all the trees came down, they never grew back.

"What happened to them?" I asked her, both cautious and fearful.

"They're gone," she said. "They're just gone."

4

Obelisks of Time

The third time I saw Washington Pass, I was headed away from the Methow Valley toward the Pacific Ocean on the curling silver thread of Highway 20. Kim was busy working on the ranch. He'd emailed me an official invite to Sheep Camp that winter. Bird was preparing to attend a sales conference for a ski brand he represented. He had plans for us that winter—rendezvous in the Cascades, the Alps, and Alaska if I happened to find myself there. Winter would be in Washington soon; trace snow dusted the summits around Twisp a week before... *it would be birdfect!*

I was lonely without my guides, and my mind wandered to another forest, an old-growth northern hardwood stand on the eastern escarpment of the Catskill Mountains. The Catskills sit on the eastern edge of the Allegheny Plateau, which starts in West Virginia and collapses, grandly, into the Hudson River a hundred miles north of New York City. Long runs of shale and sandstone break through the mountainsides, creating mile-long cliff faces hundreds of feet tall. Rivers and *kills*—an old Dutch word that means "creek"—cut gorges through the soft sedimentary rock, forming hundreds of waterfalls, swimming holes, and riverine landscapes so dramatic that Thomas Cole and Frederick Church based one of America's first homegrown painting movements on them in their Hudson River School. The second-tallest waterfall in New York (after Niagara) can be found here. As can more peaks above thirty-five hundred feet than in the entire state of Vermont.

Even more stunning to Sara and me, when we discovered this micro-range years ago, was the fact that it sits just two hours north of our Brooklyn apartment. I had always imagined that the shadow of America's largest city stretched hundreds of miles inland and that these nether regions had little to offer other than an easy commute. This is not true, as we learned on our first Catskills outing. We hiked up sandstone-topped mountains and walked through first-growth woods wilder than anything I'd seen growing up in Maine. We visited the area for a couple of years and eventually bought a wooded parcel at the foot of Overlook Mountain. We closed on the land on a Saturday morning and dragged a camping trailer onto a small promontory that night. Coyotes howled in the pitch dark long after we went to bed. Three species of owls hooted until dawn. The next morning, Sara pointed over my shoulder as we ate breakfast—at a 400-pound black bear wading across our stream.

We had lived in the steel and concrete morass of New York City for almost two decades—in tiny, expensive apartments. The existence of so much undeveloped land so nearby seemed impossible. Bobcats and fishers came down the mountain at night to prowl our woods. White-tailed deer walked silently between the trees, unflinching and uninterested in us. Red-shouldered hawks and peregrine falcons shrieked high above as eagles and crows dive-bombed them. When spring arrived two months later, a dozen puppy-sized fawns ran knock-kneed through the woods.

A thirty-foot waterfall had pounded out a gorge the size of two football fields on the northeastern edge of the plot. Striated, fifteen-foot-tall slate ledges—ancient seabed thrust up through the ground—

ran north-south through stands of hundred-foot black oak, hemlock, and maple trees. The neighborhood shared a mountain spring a half mile down the road. Our neighbor, who spent half the summer sitting in his driveway with a rifle across his lap—"waiting for the bear"—said that the spring had never stopped flowing in the forty years he'd lived there. "Hundred degrees, forty-below nights, I never seen it run dry," he said.

What happened to us in the mountains over a few years of weekend visits was a recalibration. Living in the city for so long with central heating, a garbage chute, food delivery, internet, television, public transportation, and a frighteningly complex septic system, we forgot about the mechanics of life—like where water comes from, how food is grown, what creates electricity. The cabin we built was off-grid, so these questions became very real quite quickly. We fetched water from the spring, rotated the compost toilet daily, split wood to heat the place, ferried garbage and recycling to the town dump, and wedged our phones into an east-facing window to get a cell signal. We became intimate with the water cycle, bathing in snowmelt and rainwater in a 200-gallon livestock trough fed by stream water and plunging into spring-fed swimming holes in the summer. Food came from a half dozen farms around us. If the six solar panels on the roof didn't produce enough electricity to keep Grey's night-light going, I wandered into the yard in the middle of the night to start up a backup generator. If the fire went out at two in the morning on a cold January night, I rebuilt it and sat with a book until it was hot enough to stay lit until morning.

We witnessed the vast distance that exists between most people and the environment they live in—when we occasionally rented the cabin. One winter night, a middle-aged man from Manhattan split a single piece of firewood into a thousand toothpick-sized splinters to make the woodstove burn hotter and keep his wife and toddler warmer. (It didn't.) After advising another couple to use hot coals from the last fire to ignite a new fire, the husband asked me, "Where do you keep the coal?" One guest spent an afternoon climbing a tree, tea kettle in hand, to refill a solar shower bag that was suspended with a perfectly functioning rope and pulley. After a man in his twenties inquired how to light the propane cooktop, I asked if he had a stove in his apartment. "Do we have a stove at the apartment?" he asked his friend. Once he got it lit, he asked how to turn it off.

It's impossible to fully comprehend climate change, its causes, and the coming doom it promises without knowing where your cooking gas comes from or where your garbage goes. In the fog of modern convenience—and perhaps a decade from the point of no return—the nations of the world are now carbonizing the atmosphere ten times faster than at any time in history. Power plants, consumer vehicles, inefficient buildings, irresponsible farming, and growing natural feedbacks have jacked the amount of carbon dioxide in the atmosphere to 421 parts per million. (Preindustrial carbon levels hovered around 280 parts per million.) The last time there was that much carbon dioxide in the air was three million years ago, when there were no humans on earth, sea levels were fifty feet higher, and forests grew in Antarctica. Even with the Paris Agreement and myriad other climate accords, we humans have emitted more carbon dioxide in the last ten years than we did in the entire breadth of human history up to the 1960s. (Emissions in 2019 were forty-three billion metric tons, a new record.) These high levels will be here for a while. More than half the carbon dioxide we put into the atmosphere will still be there a thousand years from now—affecting at least thirty human generations. A third of that carbon dioxide will still exist in twenty thousand years. Even with a new climate-friendly US administration and aggressive decarbonization by the rest of the world, the same computer models that have successfully projected global warming for a half century now suggest that average global temperatures could peak at not three, but *five degrees*. To keep below the once-standard safety threshold of two degrees of warming, the UN's "Emissions Gap" says, the world will have to increase emissions reduction targets—many of which are not being met now—by a factor of three.

People in the Methow Valley live a bit closer to nature and farther from convenience. Golden hayfields and buck-and-rail fences separated the road from the crumbling Twisp Formation as I drove west. Water in the Methow River was low, exposing boulders, bleached tree trunks, and long deltas of silvery river rocks. The trees grew tighter and thicker the higher I climbed. Twenty feet off the shoulder, pallet-sized bundles of deadfall, collected by Forest Service workers to prevent wildfires, sat between the pines. Sunlight splintered across the windshield, and a few wispy clouds tailed off the tallest peaks. I passed the summit of Liberty Bell and an outrageous line that Bird had skied down a few winters before. Then I was through Rainy Pass, on to the west side, free-falling through Mount Baker-Snoqualmie National Forest and the green carpet of old-growth evergreen blanketing the western Cascades.

Few ecological boundaries are as dramatic as the divide between the eastern and western Cascades. After my drive through the arid western forests and tablelands of Central Washington, the west side looked like a Tolkien jungle. Old man's beard lichens hung from trees six feet around at the trunk. Temperate rain forests, home to some of the most prodigious precipitation in the Western Hemisphere, shaded the highway. There were no bare hillsides like in Twisp; everything was layered with thick, green flora. The region supports more life by weight than does any other ecosystem in the world: massive stands of cedar, noble fir, Sitka spruce, Douglas fir, redwood, sugar pine, hemlock, sequoia, and black cottonwood. For the first time in recent history, the trees there had also been fuel for unprecedented forest fires.

The landscape on the west side is typically too wet to burn, and yet half the trees near the Ross Lake Dam were black instead of green. Wide swaths of scorched forest reached up the mountainsides. Light shone through ridgelines where the canopy had burned away. Most of the tree trunks growing close to the highway were untouched; the fire crowned most of the way up the valley. I learned later that it had burned in 2015—the same year the Cascades suffered a major snow drought, receiving 20 percent of the range's typical snowpack.

The first town on the western side looked like it had survived the blaze. A sign announcing Burn Ban in Effect leaned against a telephone pole near the Marblemount Community Hall. Hungry, and a little lost, I pulled into Chom's General Store to get a sandwich. If the name doesn't draw you in, the characters at the cash register might. What looked like a squirrel trapped in a pair of jeans and a flannel shirt spoke so frantically to the cashier that it took a minute to figure out what he was saying. There was a conspiracy afoot, something to do with the president and Congress and deep-state bad actors. Certain unnamed players in Washington, D.C., were up to something. A certain president should not go outdoors for at least twelve weeks. "Watch for an assassination attempt the day after Thanksgiving," the squirrel said. "Then the stock market dives the first week of December. Three days after Christmas, they'll start taking our guns."

The middle-aged woman behind the register nodded and twirled a pastry in her hand. She seemed to have some experience with this character and was mildly entertained by him. No one noticed me walk in, so I eavesdropped while browsing the camping section, the pesticide section, and the surprisingly well-stocked "Nuts 'n' Bolts" aisle. When talk of what the Constitution means to them came up, I grabbed a sandwich from the refrigerator and escaped the swelling tide of paranoia at Chom's to find Jon Riedel.

I had gotten to know the idea of Dr. Riedel long before I met the

Not to obsess over the "Nuts 'n' Bolts" aisle further, but, I mean, Chom's is a small store. *And this aisle has pretty much everything you need to build a house. This is the toolbox of Manifest Destiny; it's the Way of the West. Since the 1800s, these little general stores have served remote populations that can't get to a larger town. The offerings are precise; the selection vast!*

man. The story I was following led from wildfires to vanishing snow to a warming planet and a host of characters studying what the earth will look like when winter is gone. The best way to prognosticate future climates is to look at the past, including the last great winter that covered most of North America in ice several miles thick. Nowhere in the contiguous United States is there a better place to find year-round ice than in the North Cascades. More than seven hundred glaciers hang from the peaks there, many of which were first measured and recorded by Jon Riedel.

I couldn't find a direct phone number for Riedel, so I called the North Cascades National Park Service HQ and spoke with a friendly woman who did not seem to know who or what I was talking about. Then I tried the park's Glacier Monitoring Program manager, then the data manager, an ecologist, and two physical-science technicians. No one answered their phone, so I left messages and sent a round of emails. A week later, two people got back to me but had no contact info for Riedel. Another who called a few days later seemed genuinely surprised that I thought I would find Riedel at a desk or anywhere near a phone. When one of the technicians eventually gave me Riedel's email address, I broke down and asked if the man actually existed. "This is not an uncommon situation," he said and hung up.

Riedel is well known for founding the Glacier Monitoring Program in North Cascades National Park in 1993, creating a baseline from which melting and warming in the Cascades and beyond can be compared with today. (The project has since expanded to several other national parks.) Riedel's work, though, and the lore surrounding it and him, is larger than that. He had explored the most remote corners of the incredibly remote North Cascades, including places that had seen no more than a handful of human footprints. He had unearthed fossil wood from ancient forests on the flanks of Mount Baker—to gauge when they were knocked over and thus calculate glacial extent at the time and a general understanding of the climate then. He was not the stick-in-the-mud, killjoy park official I sometimes encountered on public lands, nor was he the stony-eyed scientist who spoke about the end of the world without emotion. He was a spiritual being who had a government job, had been swept away by the storyline of ancient winters, and had dedicated his life to understanding them.

I pieced much of this portrait together from a video the National Park Service had produced in 2013. In it, Riedel appears as a tiny dot walking across a glacier the size of a small New England town. He is wearing glacier glasses, a harness, standard park-issue greens, and an NPS baseball hat. He often shares the screen with students, who probe the ice behind him with long rods, searching for crevasses, melt holes, and deeper, older layers of ice. Epic pans of the Cascades show some of the tallest peaks in the Northwest draped in snow and ice, with a voiceover from Riedel reading from a book he wrote about glaciers: "The rhythms of our lives may seem weak, especially in the din of the city. In these mountains, Earth's rhythm is strong..."

The book is titled *Ice Blue Legacy: Glaciers and Glacial Landscapes of the North Cascades*. It has yet to be published, but from snippets I've seen, the writing is occasionally reminiscent of the grandiose prose of Henry David Thoreau. Glaciers are "obelisks of time" and "elders of the landscape...strung out like pearls on the backs of the mountains." Climate change is "the heat of human consumption," and ice ages "dominate our rhythm at one hundred thousand-year intervals."

The rhythm of the Cascade and Skagit Rivers was slow and steady when I headed over a steel bridge on Cascade River Road toward Riedel's house. A final round of emails had produced his cell phone number, an invitation, and directions to his home. The road narrowed and the forest crowded the soft shoulder. A mile later, it grew smaller still and angled east and north. Barbed-wire fencing lined a green pasture. The pyramidal peak of Lookout Mountain hung above the treetops.

After two wrong turns, I spotted a perfectly manicured lawn and a cedar-sided home. Riedel waited on the stoop, offering the presidential smile and wave of a lifetime national park employee. He was presidential in other ways: a slow and careful cadence to his speech; a six-foot-one frame; a perfect, unmoving part in his blond hair.

He waved me into the post-and-beam home he designed and built in 1993 with his wife, Sarah. Sarah also worked for the park, and the couple raised their two daughters there. (Both girls have names derived from local mountains.) It took two years to find the right spot—avoid the river and flooding; the landslides off the mountains; the shady side of the valley, where you don't see the sun for three months of the year; the high-tension power lines running from the Ross Lake Dam to City Light in Seattle. He built the home in a year. "The longest year of my life," he said.

Riedel was not born in the mountains. He grew up a flatlander in central Wisconsin and received his master's degree in physical geography at the University of Wisconsin–Madison in 1987. After arriving in the North Cascades in 1980, as part of the Student Conservation Association, he had to learn an entirely new way of moving—using crampons, ice axes, Prusiks, and harnesses on alpine glaciers. He learned how to rope up when walking on ice and how to rig a Z-system to pull colleagues out of a crevasse should they be unlucky enough to collapse a snow bridge and plunge in.

He wrote his master's thesis on Redoubt Glacier, set on the remote Mount Redoubt near the Canadian border. He visited the glacier for days at a time to study the history of ice there. For a young geologist, the North Cascades were a playground. Two million acres of contiguous wilderness, including the park, not only had been formed by glaciers and tectonic events but was still being formed. While Riedel was driving west to start working in the Cascades, Mount St. Helens erupted. He remembers watching fantastic yellow-purple sunrises and sunsets as he sped across the Great Plains, then seeing ash from the eruption falling from the sky. He got as close as he could to the source, but police had roped off most of the roads. He managed to see mudflows and lava flows and, later that summer on the job, glacial outburst floods and snow avalanches that plowed over entire forests.

Measuring the pace of an object that moves so incrementally that its name is synonymous for *slow* can be a monotonous task. But pace depends on perspective, and in the realm of deep time, glaciers are devastatingly dynamic. They dug the Great Lakes a quarter mile deep, flattened the Northern Plains, and shaped most major river systems and mountain ranges in North America. Glaciers carved the coasts as well and pushed till hundreds of miles, creating the foundations for entire ecosystems across the planet. If you could watch two million years of the earth's history in fast-forward, ice's dominion over terra firma—think waves crashing on a shoreline—would be terrifying.

By the 2000s, it was clear that ice in the Cascades did not have long to live. Glaciers across the region's parks, from the Olympic Peninsula to the North Cascades, had always reacted differently to various climate events. By 2007, every glacier in the park was moving in unison for the first time: backward. The rate of recession increased as the decade wore on—radically on some glaciers that had hardly moved in a century. Even with an increase in precipitation during some winters, warm air over the region continued to melt the ice. "One of the most striking things you see are lakes now where there used to be glaciers," Riedel said. "They create these depressions in the side of the mountains as they advance, then, as the glaciers melt, lakes form under them and the glacier eventually collapses into the lake."

Riedel led me to a covered porch and two rocking chairs. Winter was still a month away, but there was a chill in the air. He and Sarah spent most of their time in Anacortes, Washington, now, where their girls had gone to school. The Marblemount house was furnished but otherwise empty. No photos on the refrigerator door, no dishes in the sink, no mail stacked on the kitchen table.

Riedel leaned back in his chair and listed the status of each glacier he monitored. Most were listed among the fastest-melting glaciers in the world. It takes a massive climatic change to sync up the flow of billions of tons of ice, and to do it in such a short timeline—in Riedel's lifetime—was unthinkable. The ramifications were many. For hundreds of millions of years, ice and snow regulated the earth's climate by reflecting 80 percent of solar radiation that hit them back into space. When they melt, the exposed dark ground and water absorb sunlight and radically increase warming. Scientists call a surface's ability to reflect solar radiation its *albedo,* from the Latin term *albus,* meaning "white" or "whiteness."* The melting glaciers were causing a rapid feedback loop of diminishing albedo. There are other feedback loops too, such as how a lack of sea ice in the Arctic changes weather patterns around the world and how trillions of gallons of water from melting land ice can block major ocean currents like the Gulf Stream—altering the climate of entire continents while raising sea levels.

I had told the story of skiing and winter hundreds of times, and even the future of snow in a warmer world, but I had somehow not connected all the dots until a few years ago. We know it is going to get warmer. We know that part or all of the cryosphere will melt. But what does that actually mean to places far from a ski slope—in Los Angeles? Paris? New Delhi? My home in New York? Things don't just happen in nature. Cascades happen as interrelated systems crash, in the way that a single falling stone starts an avalanche. Melting the cryosphere was like dropping a boulder into that avalanche field. Things start to fall with it. "It's a pretty wicked series of things that all sort of tie together, and then throw in a tipping point or thresholds on top of that," Riedel said. "And you've got to believe we're approaching, whether it's the Arctic losing its sea ice or

* If the planet were completely covered in snow, the earth's albedo would drop the mean temperature from fifty-nine degrees Fahrenheit, where it sits now, to minus forty.

the Greenland Ice Sheet disappearing, many of these feedback mechanisms that could accelerate climate change."

Another way to envision the coming transformation, Riedel said, is to think of it in terms of seasons. The Northern Hemisphere enters winter when the planet tips back and half of its land surface is covered with snow and ice. Since the 1800s, two centuries of industrialization have erased thousands of years of deep freeze and created an antithetical trend—an epic summer that is overtaking winter. "It's all tied to the seasons," he said. "What we're seeing now is the growth of the summer season—and the fire season—so the forests are drying out earlier, again, due to loss of snow but also due to warmer air temperatures. The loss of snow affects the water supply, but it also affects the ecology in so many ways—not just forest fires but the growing season, pollination, hydropower, freshwater supply, river habitat, and myriad other effects because climate affects everything."

This has happened before. "Hothouse earth" three million years ago saw palm trees, giant beavers, and camels living quite happily in the Arctic. The difference being that those climate shifts were kicked off by natural forces like volcanoes and giant methane releases that typically took thousands of years to play out. The speed of current climate change has been seen only a few times in the past, and the effects will be drastic, Riedel said. Another interesting point, he added, was how we have oriented ourselves, placed our cities, established our diet and agricultural practices rigidly within the parameters of the current climate—which will no longer be here in a few decades. Riedel was saying this: the snow and ice we both loved and had explored for most of our lives was *more* than a boulder teetering on top of a mountain. It was the mountain.

Riedel has spent much of his career re-creating the climate of the last Ice Age. He and his team have walked the Cascades and studied the same Mount Adams forests that Swensen and Gleason work in. They have searched the Olympic Mountains for macrofossils. By inspecting glacial deposition zones, analyzing volcanic ash and radiocarbon-dating fossilized tree rings, pine cones, and seeds buried by a succession of advancing glaciers, they have mapped the timing of glacial advances, and thus the climate, of the last Ice Age. Riedel and his team found that climate change has had huge implications for the landscape, even during Holocene periods that marked the end of the Ice Age.

"About eight thousand years ago was the Holocene thermal maximum," he said. "That's when our summers were warmest because we were tilted more toward the sun. And at that time, there were forests growing way up on the sides of the mountains here. When it started to get colder and wetter, the glaciers started advancing and burying those forests under a pile of rubble. Then they would recede and then advance even further, with each deposit going over the top of the previous one. We found tree samples and sent them to a tree ring lab, and looking at the growth of mountain hemlock, we reconstructed glacial mass balance."*

Models based on Riedel's mass-balance measurements and present rates of melting suggest the end of most glaciers in the Pacific Northwest in a hundred years. Scientists in Europe predict a similar timescale for the Alps—replacing the iconic white summits there with arid brown peaks. The great glaciers of the Himalayas, on which I skied twenty years ago, are not much better off; most will be gone in the next century. The last ice on earth will likely remain at the poles, but even there, on the 2-mile-thick Greenland Ice Sheet and the 5.5-million-square-mile Antarctic Ice Sheet, the great white buffer that has regulated our climate for millions of years is melting or calving into the ocean at a historic rate.

---

* Glacial mass balance works like a checking account. Snow accumulation is money coming in. Melt is money going out. The mass balance is the annual sum of the two and shows if a glacier is growing or shrinking.

Leonardo da Vinci studied mountains like Mount Adams. You can see them in the background of the Mona Lisa *and* Virgin and Child with St. Anne *and in* Snow-Capped Peaks*—the latter a red chalk with white heightening on paper, now owned by Queen Elizabeth II. He sketched peaks constantly and laid the groundwork for geology, geomorphology, and orography with his writings on mountains and what made them. His greatest fascination: the power and energy of falling water that poured from them.*

Riedel was facing a different time horizon. He was approaching retirement age. Reflecting on his early days in the park and the wonder that it had sparked in him, he seemed to be in a state of disbelief that his four-decade run was over. The subject he'd been studying certainly wasn't. Quite the opposite: the Anthropocene was just getting started.

"Part of the motivation to study this is personal, for my kids," he said. "The North Cascades will see some big effects from climate change, because of our glaciers and summer water supplies. Forest fires and this place that my family values quite a bit is part of that equation, too. Are they going to be able to enjoy this place? Their lives are going to be impacted more than mine, and my grandchildren will see a bigger impact than my children, and so it'll go for generations."

After an hour on the porch, Riedel suggested that we go inside to warm up. He showed me wooden trim inside the house that he had carved into alpine scenes. The house was fifteen minutes from the park and two hours from anything else. His daughters grew up hiking, climbing, and Nordic skiing on mountains we could see through the window. When the girls went to school in Anacortes, Riedel rented out the house and lived in a shed in the backyard on weekdays.

It is in this shed that one gets to know the embodiment of Dr. Riedel. The building is the size of a large bathroom and is set in the middle of an alpine garden that he has cultivated for twenty years. The shed has a translucent lean-to roof, a cot, and an attached greenhouse with oversize windows. The siding around one window is scrolled with mountain peaks, wavy lines representing water below, and stars above. Beneath the shed roof is an image of the sun with twisted tree limbs for rays.

A stack of cordwood sits near the entrance, above which two dozen Nordic skis hang in the rafters. Outside the shed door is a fire ring that Riedel cooked his dinner on for many years and that he maintains as an homage to his days as a mountain hermit. A large vegetable garden sits just downhill. Uphill are a half dozen apple trees and two pear trees with gnarled branches that Jon had grafted "to see what would happen."

Riedel moved slowly from tree to tree, lamenting how animals and the changing climate had not been kind to them. A hundred years ago, winters in Marblemount saw 9 to 10 feet of snow. Winter minimum temperatures above 4,000 feet in the region had risen five degrees Fahrenheit since the 1950s, and the mean winter freezing elevation had risen 650 vertical feet. By the time his daughters turn sixty, he said, another five degrees Fahrenheit of warming was expected. In the past, Riedel had to wait until May to access snow-bound study sites. Now he could hike there in April, or even March. Fall was the same. The first frost arrived in Marblemount a month later than it did thirty years ago. It was the same across the United States: all six thousand weather stations in the nation's network report shorter winters and longer summers—including Juneau, Alaska, where winter has shrunk by thirty-two days, and Los Angeles, where it has declined by fifty-two days.

What the public—and I, in the not-so-distant past—didn't understand were the dangers that accompanied this shift in seasons. Riedel pointed to another example in his own backyard. In 1959, summer glacial runoff in the Skagit River basin—the largest watershed that feeds Puget Sound—provided about 171 billion gallons of fresh water to the river, forests, lakes, salmon populations, ecosystems, and communities. In 2013, that number shrank to 129 billion—a 24 percent decrease. The loss equaled about a hundred-year supply of water to Skagit County at the current annual use rate. There is no plan to replace this glacial runoff with another source of water.

The sun disappeared behind the mountains as Riedel showed me out. There was no wind, and the orchard and garden reflected shades of brown in the dying light. Riedel gave me his presidential wave from the stoop as I drove away. I found my way back to Cascade River Road, the steel bridge, and Highway 20, which would deliver me to Seattle. There I would catch a red-eye to New York, walk into my home the next morning, hug Sara and Grey, and take my notebooks and computer to the office—where I'd sit for a week trying to make some sense of it all.

A United Nations report published in 2019 shed some light. It was the first official Intergovernmental Panel on Climate Change

(IPCC) study to focus on the melting cryosphere. The timing was auspicious; winter seemed to be unraveling sooner than expected. The document opens with the line "All people on Earth depend directly or indirectly on the ocean and cryosphere." It closes with a warning that we are rapidly approaching irreversible thresholds for 670 million people who live in snowy, high mountain regions. Everyone living below them was not far behind. In all these areas, in all parts of the world, the report found, "the depth, extent and duration of snow cover had declined over recent decades."

I thought about remote mountains in Alaska, glaciers in Europe, tiny hills I skied as a kid. I thought about Kim and Bird and that first day on Washington Pass, where the nascent threads of this story connected fires, forest, snow, and winter. Scientists and activists have pointed out our effect on nature for so long that it was a platitude to say all things on earth were connected. And yet that was my awakening. That and the fact that *right now* is not as the world has always been. *Right now* is a tiny sliver of time and climate created by a fortuitous confluence of natural forces and astronomical events—like the little nudge Earth received billions of years ago, knocking it off its axis, and the cadence of the ice ages and interglacial periods. What was even more obvious, and frightening, was that the winter season was slipping away. Subzero January nights, the first frost of fall, layers of fresh snow, icy peaks hanging above our heads. Without winter as our buttress against the power of the sun, many more thresholds are about to be crossed, unleashing untold natural disasters. Not in the centuries to come, not even at the end of this century, but right now.

5

The Final Questions

I couldn't see the Skagit River from Highway 20, but I could hear it if I stopped the car, which I did, several times, to listen to the dull roar of glacial melt tumbling toward the Pacific. The view through the windshield was a singular stratum of old-growth forest, prairie, field, and river. Hundred-foot spruces grew right to the pavement, except where road crews had cut them back to protect telephone lines, transfer stations, bridges, and access roads. Every few miles, a kitschy hotel—the kind that often exists on the periphery of national parks—popped up. The Glacier Peak Resort & Eatery, with its wagon wheel thresholds and taxidermic décor, had a panoramic view of Glacier Peak so immediate and so unobstructed, the mountain looked like a hazard. Down the road was the Totem Trail Motel, an old-school motor lodge where you could get a double room for the price of a coffee and a sandwich in Brooklyn.

After all the interviews, revelations, confessions, treks, bonfires, meals, and contemplative road trips during which I tried, mostly in vain, to process everything, the drive to the airport seemed like a telepathic journey from a future catastrophe to a woefully naive present. Grey and Sara were likely practicing bedtime rituals right then—teeth brushing, milk drinking, stuffed-animal spooning. How was I supposed to explain this to them? Where exactly should I drop this flaming bag of apocalyptic dog feces? On my friends? My editor? You, the poor reader in your climate-controlled home, kicking back with what you thought was a rip-roaring adventure story? How was I supposed to go about my breezy routine of bicycle commuting, food-truck tasting, and backyard gardening with this nightmare sitting on my shoulder?

All the mountains I'd skied had always been distinct in my mind. Now I saw them knitted together in a white blanket. Knowing that a winter-less, inhospitable climate was coming for them and for Sara and me was one thing. But the thought of Grey scratching out an existence in a superheated world was sickening. What would she do in the year 2080, when most of the snow in the United States would be gone and New York City's climate would be that of Jonesboro, Arkansas, today? How could anyone think about anything else right now? If I were diagnosed with stage four cancer, I would do what the doctor prescribed. So when tens of thousands of the planet's doctors tell us that we are cooking the earth in an atmospheric turkey bag—explaining how, in a few short decades, snow, winter, the cadence of seasons, water availability, the world economy, and even civilization as we know them will all hit a steep nosedive—why have most citizens of the world ignored the doctors' orders?

I branched off the North Cascades Highway onto Route 530 at a closed, dystopic gas station—where trees grew through the pavement and shrubs covered the pumps. The ever-widening Skagit River rushed below a concrete bridge as I entered the Sauk Prairie, a wedge of farm and timberland set between the Sauk and North Fork Stillaguamish Rivers. An hour later, a succession of voluminous red barns, high-tension power lines, and the Outback Kangaroo Farm marked the outskirts of Arlington—the last agricultural outpost on Route 530 and Kelly Gleason's childhood exile.

I imagined Gleason walking Arlington's streets as a rebellious teenager—ripped jeans, 1980s ski jacket, skipping school to catch a ride to Mount Baker. I had been kicking myself for not asking Gleason a few final questions while we were on Mount Adams. It is easy to get caught up in the mellifluous cadence of research-speak and ultimately lose sight of what it means to our daily lives. The questions had been bugging me for some time, and writing this now, I realize that they might be bugging you too. They may well be the reason you picked up this book, and so, for both of our sakes, I called Gleason and asked:

How many winters do we have left?

Will they come back?

What will replace them?

What will life be like without winter?

"Yeah, I'm not ready to deal with all that," Gleason said. It wasn't a punt. It was honest and pretty close to the answer I expected. Gleason is hardened. She does not recognize fear the same way I do. (Remember that she wants her final moments to be set in the jaws of a mountain lion.) She was born with the exact skills a scientist needs, and like every scientist I had met, she resisted drawing the obvious conclusion from her work.

"I am a total pessimist," she said. "Generally skeptical, but I don't think I'm quite ready to accept it, you know? I mean, we already have had these years in the Northwest like 2013, 2014, 2015. Those were essentially years without winter. Leaves fell off the trees, but there was no snow. Our modeling experiments, at most, go to 2100. At this point, keeping warming to two degrees Celsius is just laughable, right? I mean, we're at least going to hit four."

I had not considered four degrees of warming. The IPCC predicts that anything over one and a half degrees Celsius will be an unmitigated global disaster. Imagine Kim Maltais's torched backyard spread out across the US West. Four degrees was more on the order of mass extinction, food and water scarcity, political instability, and sea level rise of almost thirty feet, reshaping the world's continents. Snow would almost assuredly be gone except seasonally on a few high-altitude ranges.

"The entire Pacific Northwest and the Sierras?" Gleason said. "That snow is not going to be there. I mean, it's already very near zero degrees Celsius throughout the winter there. It's already really vulnerable, warm snow. We're already feeling it. All the little mom-and-pop ski shops are going out of business. Ski resorts are going out of business."*

Like Riedel, Gleason was most worried about the rivers, cities, and agricultural corridors in California and the Northwest that don't have a backup plan for losing meltwater. "Snow acts like a reservoir in the mountains, and Oregon, Washington, and the Sierras are going to see profound snow loss," she said. "It's like a huge dam keeping the water stored for us during the wet season and releasing it during a period of high demand, when hydropower, agriculture, and water districts need it. And we don't have the reservoir capacity to compensate for that loss."

Like the vanishing forests in New Mexico, some burned forests in the Northwest were also not growing back, she said. Wildfire depletes nitrogen and kills microorganisms in the soil, altering its chemical constitution and the nutrient bioavailability that helps trees and plants grow. Before the forest can regenerate, the soil needs to repair itself. But in burned forests of the West, warmer surface temperatures and a lack of shade were hindering the process,

* Other doomed winter pursuits, besides snowman-making and snowball fights, include Nordic skiing, snowmobiling, pond hockey, and the surprisingly robust ice fishing industry. The last two activities would disappear as five thousand lakes in the Northern Hemisphere permanently lose their winter ice within the next few decades.

sometimes permanently. Seventeen thousand square miles of forests in seasonal snow zones had burned since 2000, and that area was expected to swell by a factor of twenty by the end of the century. With a host of other climate-induced effects, wildfire could contribute to a trend in the West known as desertification*—which 40 percent of the US West is currently headed for.

Gleason wasn't sure about efforts by the National Forest Service to develop genetically modified trees in a kind of off-the-cuff geoengineering experiment on Western forests, and their success—and mass implementation of the plan—was doubtful. Forest Service employees had even transplanted trees to different elevations to try to restore forests, but had gotten mixed results.

"Even in Portland, people don't seem to get it," Gleason said. "All the cedar trees are dying in the city because it's warmer and drier now. People are freaking out, like, 'What's going on?' And I'm like, 'Well, what do you mean, what's going on? Haven't we figured this out already, folks?' "

I envisioned tumbleweeds blowing through the cities of the West as Americans walked the Oregon Trail, this time eastward, toward water, sustenance, safety, and employment. It was a macabre fantasy but entirely possible in the next few centuries. Mountains faded in the rearview mirror as Route 530 curved southwest; I was at the coast. The north and south forks of the Stillaguamish River merged on the north side of town. Both forks originate deep in the Cascades and deliver snowmelt and rainfall to farmland, as well as healthy runs of Chinook, coho, chum, pink, and sockeye salmon.

A coil of silver-black clouds rolled in from the Pacific. I merged with a cavalcade of SUVs and all-wheel-drive station wagons on Interstate 5, and it struck me that the last two books I had written

* Desertification is the process by which fertile land becomes desert, typically because of drought, deforestation, or inappropriate agriculture.

all started or ended right here—on the fringe of the continent, where the Cascades meet the Pacific Ocean. I wasn't sure what that meant. I was either going in circles, or something important was happening here. A power vortex of some kind. The proximity of North America's westerly tectonic fold that eats up two inches of the Pacific Plate annually. The pioneering spirit of the Northwest that now expresses itself in intricate tattoos, lush facial hair, and bold, analeptic coffee.

I headed south through the railroad capital of Everett, then Northgate, Eastlake, and the concrete arteries of I-5 as it brushes past the timber baron mansions, hipster coffee shops, and gay bars of Capitol Hill before descending to the scrubland plateau that is the approach to Sea-Tac International Airport. I had cut it close and barely made it through the miles-long maze travelers must follow from parking lot to Jetway. Forty minutes after turning in the rental car, I strapped myself into a 757 that spewed carbon and various other pollutants into the air as it climbed. At a cruising altitude of thirty-four thousand feet, where a surge of carbon dioxide now intermingles in our atmosphere, two hundred fellow travelers and I reclined our seats and flicked through channels on tiny seatback televisions.

It was pleasant on the plane. Warm. Safe-feeling, even. The sedating effect of modern convenience made it seem like everything was going to be all right, like someone would figure everything out. Gleason was not alone; I was also not ready. I knew what was coming, knew some degree of it was unstoppable, but I didn't fully accept it. Maybe there would be a technological Hail Mary. Perhaps Bill Gates or Elon Musk would save us. Maybe the planet would mend itself and our children would live through the full spectrum of seasons in a hospitable, if not ever-changing, climate. That would be nice, I thought. Then I reached for the screen and searched for a movie, a football game, a comedy, any possible distraction.

The Icefield

6

The Law of High Latitudes

We know where it was. It gouged mountainsides, dug out tarns and kettle lakes, and plowed long, winding moraines for hundreds of miles. It came from the north, advancing from the Arctic Sea, Baffin Island, and the great white mass of Greenland, eventually engulfing two-thirds of North America. New England was covered with ice several miles thick for hundreds of thousands of years. New York, Washington, Seattle, and much of the Upper Midwest were entombed by glaciers. The ice sheets that excavated the earth's surface weighted the crust down hundreds of feet—flooding coastlines and shaping oceans, fjords, the Great Lakes, and most of the major mountain ranges on the continent. The ice mass was so immense that it had its own gravitational field that raised local sea levels and affected tides.*

We know where the ice was, and we know where it is now. Most of the cryosphere can be found in northern Canada, Alaska, Russia, Greenland, and the poles, but fragments exist in other mountain ranges and highlands, even some close to the equator. To most

* Because glaciers plowed so much mass—which creates gravity—away from the Hudson Bay region in Canada, people there weigh slightly less, astonishingly, than they would anywhere else.

citizens on earth, this frozen landscape—which can see temperatures dip to minus 120 degrees Fahrenheit—is an alien place best avoided unless you are wearing skis or driving a dogsled. On the relief globe that I used to consult, it was negative white space: no cities, towns, or even life seemed to exist there. The white splotches were an antipode to the steaming green belt of vegetation circling the middle of the planet.

Snowscapes in high latitudes abide by a different set of natural laws. Ice and snow absorb red and yellow bands of light, creating an eerie blue hue. Because flakes form around airborne dust particles, bacteria, and pollen, snowstorms clean the air and even electrify it as falling crystals ionize the atmosphere. A quadrillion snowflakes are made every second in the clouds; ten times the weight of the planet has fallen as snow over earth's history. Some of that snow even comes from you. Humans exhale a liter of water vapor every day. A fraction of those droplets rise into the clouds and, on a winter day, can join with up to a billion water molecules to form a single flake—which may find its way back to your jacket sleeve.

Snow and ice telegraph their spectral powers around the world. A snowy Siberia affects American weather as much as or more than El Niño does, as the albedo of the West Siberian Plain shifts atmospheric circulation over the Pacific, bringing colder winters to the United States.* The process of freezing and melting water has local cooling and warming effects in the fall and spring, buffering seasonal shifts. The cryosphere has even affected the evolution of our species. So much water was locked up in ice sheets during the ice ages, the ancient forests of East Africa dried up and were replaced

* The ancient Columbia River floods that carved the gorge outside Portland—I drove through the Columbia River Gorge on the way to Mount Adams—also affected the weather, cooling the Pacific and steering winds and storm tracks across North America.

by grasslands. Living on that arid African savannah was a branch of the hominid tree—*Homo sapiens*. This species was suddenly forced to hunt as a group, move through the Rift Valley as nomads, and adapt as lake levels fell. Around twenty thousand years ago, some of those nomads found their way into North America on an ice bridge across the Bering Strait.

These properties of the frozen world have scientists and other people asking a third question: Where is the ice going? And, I will add, how is all this relevant to the end of winter? Has the delicate balancing act among liquid, solid, and gaseous water on earth been tipped for good? Are we merely riding out the disaster we set in motion, or is there something we can do? Finally, who is in charge of figuring all this out?

One possibility was this man right here, Seth Campbell, clad in microfiber and hunched over his laptop at the century-old Alaskan Hotel and Bar in Juneau. How did Dr. Campbell find himself on the edge of one of the largest ice formations on the planet, sequestered in an old-timey haunt that, similar to the cryosphere, was a world of its own—with its carnival-like atmosphere, tobacco-stained old-timers glowering in the corner, troubadour wearing a boater and singing show tunes to tourists on the sidewalk? There was even a ghost of Alaska's past, a doe-eyed bartender named Kevin, who, with a conductor's hat and engineer overalls—one suspender coyly unfastened—wanted to know Campbell's drink order.

Stout was the answer. Kevin winked and shuffled toward the taps. I had already ordered—whiskey—an instinctual response to the panhandler vibe coursing through the room. I had flown seven thousand miles to listen to Campbell talk about the state of the cryosphere, and I needed a drink. When he had emailed me the invitation, I had been on a much-needed family vacation, sitting poolside on a record-breaking 115-degree day. Grey was learning to paddle an inflatable shark around the pool. Sara was sending my nieces into hysterics frying an egg on the slate patio. The air had felt as if an oven door had been left open. I had had a hard time imagining that a shard of ice existed on the planet right then, but Campbell's email convinced me not only that ice did exist but that an emergency was unfolding on it.

Alaska had also recently experienced record heat—ninety degrees in Anchorage on July 4—and the Juneau Icefield Research Program (JIRP) that Campbell directs was melting out. The escape route for sixty researchers and students on the icefield was blocked by deadly crevasses up to a hundred feet deep. While the early melt was unheard of at JIRP, the danger was not.* The program's founder, Maynard Malcolm Miller, practically invented field glaciology in the 1940s, and attendees had been traversing the icefield since 1950—as part of the second-oldest glacial monitoring program in the world. Dr. Miller, or "M3," was a bit of a loose cannon in life and in the scientific community. He loathed the politics, the snail's pace of academia, and the closed-mindedness of many of his colleagues. On the ice he was a wild man, equipped with a wood-handled ice axe, a heavy climbing rope around his neck, and a mischievous twinkle in his eye. He did not appreciate students who expressed fear, weakness, or anything but wild exuberance at being invited to the program. If campers didn't write their name on a piece of gear as instructed, M3 would write "Maynard" on it and keep it to teach them a lesson.

In photos from the 1950s, he has the slightly foppish look of an alpine Jacques Cousteau—signature white newsboy cap, trimmed beard, down parka.† He typically carried a small rucksack and had

---

* Campbell's current position at the University of Maine, another pioneering institution in Arctic research, was to help fill the role that Gordon Hamilton, a prominent climate scientist, held until he died after falling into a crevasse in Antarctica.

† Imagine Bill Murray playing the lead character in *The Life Aquatic with Steve Zissou* to get what I mean. Zissou was an amalgamation of Jacques Cousteau, the character of Guido Anselmi played by Marcello Mastroianni from the Federico Fellini film *8½*, and Murray himself.

Maynard Malcolm Miller, fondly nicknamed M3. (Photo courtesy of Juneau Icefield Research Program)

his arm slung around a colleague. He grew up hiking the Cascades around Tacoma, Washington, and served on a navy destroyer in eleven active campaigns in World War II before being reassigned to the Arctic to study sea ice—specifically, how it affects the trajectory of missiles pointed at the USSR. He earned degrees in geology from Columbia, Cambridge, and Harvard, serving as Mountaineering Club president at Harvard, where he designed masochistic, far-flung expeditions for his unwitting Ivy League brothers. In 1946, when M3 was twenty-five years old, he was assigned to research the

Juneau Icefield. A few years later, he brought students along and quietly launched a teaching program under the umbrella of "military research"—and military funding.*

Even on a good year, when crevasses are safely covered, the JIRP experience can be somewhat hellish. The 1,500-square-mile icefield is the fifth-largest ice mass in North America. JIRPers cross it, from Juneau to Atlin, British Columbia, over eight weeks on touring skis, carrying fifty- to eighty-pound packs. When the sun is out, the brightness is overwhelming. Solar radiation streams from above and reflects from below so powerfully that students are required to wear glacier glasses to avoid searing their retinas and developing snow blindness. The rest of the time, it rains, constantly. It is always cold, except for days when the sun burns through and the snow gets so soft you sink to your knees with every step. Students are roped together much of the time and learn to self-arrest with an ice axe so they don't fall down a crevasse or off a thousand-foot cliff. On bare ice, they learn to walk alone with crampons strapped to their boots. Students spend weeks measuring invisible things like glacier surface velocity, mass balance, snowmelt, percolation, ablation, evaporation, sublimation—meticulously recording the data in notebooks, then transferring it to digital files. (There are no evening beers at JIRP; it is a dry campus.)

Young, brilliant minds are not always attached to young, athletic bodies, so tears and breakdowns are common. This is part of the founding principle of the camp—to teach young scientists to endure the rigors of endless research in a freezing laboratory with little communication with the outside world. Another is to have students

---

* America's first Mount Everest Expedition chose M3 to be chief scientist in 1963. On the expedition, he famously offered up vials of meltwater he'd meticulously collected on the Khumbu Icefall to delirious climbers on their way down from the summit, likely saving a few lives.

from all over the world live, cook, study, sing, crap, nap, and travel together on the ice they study, engendering a bond among one another and it. The program seems to have worked. There is much joy—or *JIRPness,* as campers call it—in the camp. In a crowd of fifty to seventy somewhat introverted scientists, who ordinarily spend hundreds of hours at home alone in a lab but who now are communing with peers and mentors on the very ice and rock they have been analyzing, JIRPness can amplify to dangerous levels.

The program is no easier on organizers and faculty like Campbell and his colleagues. Students and teachers sleep and eat in the same ramshackle shelters and mess halls. (JIRP has thirty-two camps spread across the icefield, though only a dozen are regularly used.) For most attendees, the program is worth it. Over the last seventy-four years, JIRP has hosted thousands of young geoscientists—many of whom have gone on to administer or work on such world-changing administrations as NASA's Mars Rover Missions, Barack Obama's climate policy team, pivotal IPCC assessment reports, and cutting-edge climate labs and university programs around the world.* Hundreds of signatures from JIRP alumni on the beams and walls of camps now represent multiple generations of the earth science community.

M3 was just twenty-eight years old when he set the course for JIRP. In his first few years on the icefield, he noticed a consistent trend in glacial mass-balance calculations and made one of the first connections between climate change and glacial retreat. "Tremendous recession of ice is going on; the Earth definitely seems to be warming up," he wrote in a 1949 study published in *Science*

* One such alum on JIRP's board is MacArthur Fellow Ben Santer, who wrote the first scientific consensus that humans were changing the climate in the 1995 IPCC report. In 2020, climate deniers in the federal government pulled Dr. Santer's funding and shut down his research at the Lawrence Livermore National Lab.

Illustrated. "These glaciers are telling a story. Only when we learn to read it can we know whether the Earth is warming up on a major scale."

The Milankovitch Cycle

I'd heard about Campbell from one of my high school ski buddies. Campbell was a fellow Mainer and, I would soon learn, a Han Solo–esque ice hustler who had been on more than sixty polar expeditions over the last few decades. He spends eighteen hours a day writing grants and quarterbacking logistics to get scientists onto glaciers and ice sheets, ultimately trying to figure out the final, all-important last question: what will happen if ice disappears?

Before we get to all that, you need to know two unbelievable facts about Campbell: (1) he does not sleep, and (2) he has never drunk a cup of coffee in his forty-two-year life. A three-hour nap at any point in the day pretty much does it for him. He said his father, Bill Campbell, is to blame. Old Man Campbell woke him and his sister at four in the morning every day before work so that the kids would "know what that felt like." This was in Nobleboro, Maine, a pinprick of a town about an hour northeast of Portland and two hours south of the island I grew up on—where I knew dozens of old, wizened, tough-love men exactly like Seth's dad.

Bill Campbell was an environmental scientist, employed to research how dams, power stations, and the former Maine Yankee Nuclear Power Plant in Wiscasset affected the forests, rivers, and ocean around it. Off duty, Bill was a woodsman who ran his own trapline and operated a sizable fur-trading empire out of the family's garage. He was a modern-day coureur de bois, as French-Canadian trappers called them. Campbell and his sister learned to cut pelts off hundreds of animals that their father then sent to Canada to be processed. "My sister and I joked that he should've been born in the

1800s," Campbell said. "At any given time, we'd have thousands of pelts in the garage. I remember a boyfriend of my sister's coming over and being horrified."

Bad actors and abusive parents in the neighborhood did not last long. Bill Campbell stood six feet one and brought a kind of Wild West justice to town. "I remember one kid showing up after his father had beat him up," Seth Campbell said. "Dad put me and the kid in the truck and drove over to his house. I'm not sure what happened, but I remember seeing the kid's father fly over the kitchen table. He never hit his boy again, as far as I know."

It was likely an inherited sense of rectitude that launched Seth Campbell on his career, because unlike most jobs that require employees to show up at a location, execute a certain skill over a certain period, take their paycheck, and then do it again day after day, Campbell's job as a scientist, and as JIRP director, was to create—and pay for—his position before actually doing it. In other words, half his life was spent writing grant proposals to study ice; shore up JIRP's ramshackle facilities; find and assign various scientists, researchers, instructors, managers, and support staff according to their evolving skill set; explain to everyone not only that they will be working for free but that they will also be billed for expenses; and let them know that they will be sleeping in bunk beds in unheated huts and using outhouses with a fairly persistent rat problem.

"So, this is Alaska," Campbell said, pointing to a colored grid on his laptop screen. "Arctic up here, Pacific Ocean here. This is where the Aleutian low [low-pressure system] is situated. Over a long-term climate, if you shift the Aleutian low up this way versus inland, it changes all the processes over Alaska. Another thing that we know changes: if you remove sea ice in the Arctic, it shifts weather patterns, so that's what we think is happening right now."

"Right now" is an interesting moment in the realm of earth

science. Things once thought to be static are now moving quickly. The planet is hotter than it has been for at least twelve thousand years. The poles are melting. Sea ice is vanishing. Record-breaking tropical cyclones now register *stormquakes* of up to 3.5 on the Richter scale. Heat waves in Asia, Europe, and Australia in 2019 hospitalized tens of thousands. Massive, underground "zombie fires" burn all year long in the carbon-rich peat of the Arctic. Humans have altered three-quarters of all land on earth and two-thirds of the oceans. Human activity has also killed off two-thirds of the world's wildlife population in the last fifty years, leaving a fifth of all species facing extinction in the near future.

How have glaciers reacted to warming? Similar to what Riedel found in the North Cascades, the movement of nearly every glacier *in the world* is now synchronized—backward—as they melt at a historic rate. Half of the glacial ice in the Alps has vanished since the 1800s. The Himalayas are losing eight billion tons of ice a year—enough to fill more than three million Olympic-size swimming pools. The Juneau Icefield, itself the size of Rhode Island, will likely lose 60 percent of its ice by the end of the century.* Since 1961, Alaskan glaciers have lost more ice mass than has any mountain glacier system on the planet.† As nine million square miles of permafrost thaws in the Arctic, more than a billion tons of greenhouse gases frozen within it will soon be released into the atmosphere, potentially warming the planet many times more than humans have. In sum, recent ice melt on earth, including unthinkable amounts in Green-

* Another barometer of the state of the world is a new study showing that in two and a half centuries, humans have altered the landscape of North America more than the recession of the last Ice Age has.

† A recent paper led by US Geological Survey glaciologist and JIRPer Chris McNeil shows that meltwater off the Taku Glacier between 2013 and 2018 would fill Green Bay's Lambeau Field stadium every day. The Taku is one of more than sixty thousand glaciers in Alaska.

land and Antarctica, has redistributed so much mass over the last few decades that it has altered the rotational axis of the planet.

Warming hasn't hampered the pioneering spirit of Alaska's capital city. Images of Juneau locals wearing bikinis appeared in mainstream media outlets for much of the summer. When Campbell and I exited the bar, a sample of the state's newfound joie de vivre was on display outside the historic clapboard buildings lining South Franklin Street. In the 1800s, gold had been the city's first boom, as the California gold rush dried up and the forty-niners wandered north to find their fortunes. So there were several century-old saloons to explore. The 1855 pine green J. J. Stocker building stood opposite the 1860 Imperial Billiard & Bar. Western facades and spacious sidewalks allowed tourists ample room from which to window-shop a seemingly endless succession of gift stores.

A half mile out to sea, the mother ships of Juneau's hospitality industry bobbed in the harbor—three massive cruise ships, each the size of a twenty-story office building laid on its side. Five arrive each day in Juneau, Campbell said. A temperate climate has boosted the cruising business in the extreme north—and with it, extreme gifting. Cruise lines recently bought up most of the gift shops at each of their Alaskan ports—so they could dig a bit deeper into passengers' wallets.

The trinkets packed into storefronts represent nature to most of Juneau's tourists: exotic objects to look at, purchase, and place in your pocket. Real nature—living, breathing, wild animals; dense forests; precipitous mountains; inclement weather; stinging bees!—was hardly worth the hassle. Why not jump on a week-long cruise for $775, including unlimited buffet options, sugary cocktails, and a *view* of nature through a watertight, oval cabin window? Imagine having to actually walk up one of these steep hills? Having to stand out in the rain! Just think that someday you might have to wade across a stream and get your feet wet. Or peer into the dark chasm of your own mortality during a live encounter with an undomesticated animal! (Maybe the passengers have a point. A few days after I got home, three visiting kayakers were found dead beneath the toe of the Valdez Glacier in Alaska, after it unceremoniously calved a few hundred thousand pounds of ice on top of them.)

Campbell summarized earth's climate over the last hundred million years as we walked. The world has been getting drier and cooler for around fifty million years. For the last two and a half million, it has oscillated in and out of ice ages and interglacial periods according to the Milankovitch cycle, which is driven by the planet's uneven orbit and tilt around the sun. The cycle has three parameters. The first is eccentricity, which refers to Earth's elliptical orbit. The next is obliquity, or how its axis angles away from the sun. The last parameter is precession, which references the axis's wobbling like a toy top. Add up these three measures, and you get the cause of most prehistoric climate change, except when you don't...

Warming, cooling, and the formation of ice haven't followed the Milankovitch cycle precisely. Obliquity, which runs on a 40,000-year cycle, is stronger than eccentricity, yet ice ages have followed the eccentricity rhythm. (Except when they didn't—between 2.5 million and 800,000 years ago.) A recent MIT study found that the *rate* of warming and cooling during these times influenced conditions more than the total amount of temperature change, explaining several "snowball Earth" scenarios hundreds of millions of years ago, when the planet was encased in ice. Another quandary: the solar radiation that Earth receives from the sun has increased by 30 percent since the planet was born. So why has the planet been cooling off for roughly fifty million years?

These aberrations are among the many mysteries that have recently brought Campbell and his colleagues' work into the spotlight. Just a few decades ago, it was hard to get funding for glacial research. Now, federal agencies like NASA and the National Science

Seth Campbell pointing at Camp 18. Glaciers are not an action; they are a reaction to winter. When and where the earth is below freezing, water exists as a solid. The colder and longer the winter, the higher snow and ice will stack up and the faster the glaciers will grow. Over the last two and a half million years, rivers of ice covered much of earth, shaping most topographical features on the planet, reflecting a vast amount of the sun's energy, and transforming earth's climate.

Foundation are seeking cryosphere projects to fund. They want to know how long glaciers will last and, as M3 discovered, how the effect of climate on glaciers could in fact help predict future climates. But increasingly, agencies are interested in another measure. They want to know what is trapped *inside* the ice and whether ice cores taken from glaciers and ice sheets can paint a picture of the future.

A German scientist named Johannes Georgi accidentally pioneered the science of ice-core analysis in Greenland in 1930. He had been left on the summit of the island's massive ice sheet, at 9,850 feet, to establish a monitoring station while his colleagues traveled 250 miles by dogsled to fetch supplies. Even in summer, fifty-degree temperature swings between night and day were not uncommon, so he dug down a few steps and carved a storeroom in the ice to protect sensitive weather instruments. A couple of months later, he and Ernst Sorge, a fellow scientist, dug farther down and built sleeping quarters. The temperature beneath the ice that winter was sometimes sixty degrees warmer than outside. Sorge continued the excavation for another year to a depth of just over fifty feet. He noticed the increasing density of the snow as he shoveled—and wondered if he could use the formula for density (density = mass/volume) to figure out when each stratum of snow first fell. He eventually delineated twenty-seven years of snowfall in the shaft, creating the basis for ice-core dating.

Modern ice-core drilling is far more efficient than digging a snow pit.* Drill crews date years by finding melt-freeze layers, as Sorge did. But now they find much more: ash smears from volcanic eruptions, dust, pollen, uranium, isotopes, and methanesulfonic acid. Labs extract air from bubbles frozen in the ice and measure how much carbon dioxide was in the atmosphere during some of the hottest and coldest periods in the planet's history. (All the hot periods saw significantly increased carbon dioxide.) Traces of lead in Greenland ice cores—emitted when the Romans minted silver coins—correlate with booms and busts in the empire's economy. Salt in ice

---

* JIRP still requires students to dig snow pits, partly to show them how miserable glaciology used to be.

cores reveals how big the waves hitting Antarctica were after the six-mile-wide Chicxulub meteorite hit near the Yucatán Peninsula around sixty-six million years ago. A 230-foot core from the Colle Gnifetti glacier in the Swiss-Italian Alps revealed traces of lead pollution from Middle Age British mines rising and falling according to historical events—like new taxation policies, wars, and the killing of Thomas Becket, the archbishop of Canterbury.

The oldest known ice was retrieved by a Princeton team in the Alan Hills region of Antarctica and was dated at 2.6 million years old. Scientists with the Australian Antarctic Division recently unveiled a drill that could reach nearly two miles beneath the thickest ice in Antarctica. There they hope to sample ice, air, and ancient detritus thought to be more than a million years old. Using stainless steel, aluminum, bronze, and titanium for the shaft and bit, and a 500-ton drilling base, the crew expects it will reach primordial ice around 2030. One question they hope to answer: why did the ice ages shift from a 41,000-year cycle to a 100,000-year one?

JIRP was among the first to drill ice cores, in the 1950s, and part of Campbell's mission as director was to bring drilling and a host of other tech innovations back to the program. The program needed it. Seventy-one years is a long time to maintain excellence in anything, and JIRP had begun to wither during M3's last years. In 2018, *Popular Mechanics* published an article subtitled "How a Legendary Glacier Study Almost Died When the World Needed It Most." Class size had shrunk from over a hundred students to a dozen. Fewer scientists were interested in researching and teaching there. Some facilities had fallen into disrepair. After M3 stepped down from his director role in 2010, following a sixty-four-year run—he died in 2014—JIRP's alumni stepped in with emergency funding and cycled through several directors before landing two former students as program manager and director in 2018: Annie Boucher and Seth Campbell.

By 2019, the crew of former JIRPers had the program back on track. This year, Campbell said, visiting researchers included a crew from Caltech and NASA Jet Propulsion Laboratory, who successfully tested a distributed acoustic sensing system on the ice.* Another crew, also funded by NASA, would use a spectrometer to look for signs of microbial communities living in glacial ice, while others collected rock samples to study deglaciation across the icefield and geomorphology erosion processes. It was the last crew, though, that Campbell was most excited about. They would be testing a thermo ice-core drill rig that would retrieve samples three hundred meters beneath the Matthes Glacier. Tomorrow, he said, a helicopter would fly us in to see it all. That is, if the rain let up.

Nature Remains Oblivious

The scene at JIRP headquarters that evening looked like a Marvel Comics film—in which a gang of superheroes with various abilities strategizes how to fight an invisible enemy. This was no stealth battleship anchored offshore, though. It was a converted garage with a concrete floor, a tool bench, and four oversize trash and recycling cans in the corner. Instead of high-resolution screens indicating the status of glaciers around the world, there were bookshelves screwed into the Sheetrock, each holding fifty pounds of dissertations and reports by M3 and other distinguished JIRP alumni. Twenty body-bag duffels and internal-frame packs lay in a heap by the roll gate. Maps of the Juneau Icefield taped and thumbtacked together hung from the walls. A full kitchen had been installed, as well as a large

* A glimpse into the world of cryospheric innovation is apparent in how the distributed acoustic sensing system works. It uses a kind of fiber-optic cable laid out over more than a half mile of ice to act as thousands of seismometers capturing glacier movement at very high spatial and temporal resolution. It would take a year to install that many individual seismometers, versus a couple of days for the cable.

table layered with notebooks, bags of tortilla chips, and a half dozen hot sauces.

Our heroes chatted excitedly around the table. All were JIRP faculty, and in keeping with our superhero analogy, all had highly focused and slightly bizarre skill sets. Take Donovan "Donny" Dennis, a boyishly handsome twenty-eight-year-old, sitting at the head of the table in loose-fitting long underwear, glasses, and a pompadour. His special skill—which speaks so purely to the minutiae of information, measurement, and observation in the broad concept of geology—is understanding how and when rocks fall from the sides of a valley after a glacier retreats. Next to Dennis was Allie Balter, a flaxen-haired geochemistry PhD student at Columbia University. Balter's headband, glowing smile, and staggering sense of positivity were a smokescreen. She was, in fact, a seasoned time traveler who wandered through the Pliocene and Miocene epochs on the frozen wastelands of Antarctica and Greenland—then returned home with rock and ice samples that helped colleagues determine the weather on Thanksgiving Day, nine hundred thousand years ago.

Other members of the squad included Kiya Riverman, a rock star glaciologist who got into the field after her professor noticed that she would physically fit into a glacial melt tube that he needed samples from; Michaela King, a PhD student from Ohio State who was studying how glaciers change weather patterns; and, finally, my favorite, a mysterious older fellow named Wilson Clayton, whose sparse gray beard, distant gaze, and ambiguous mission on the ice gave him a nebulous Big Lebowski aura. Wilson was officially sponsored by the Colorado School of Mines to study how surface meltwater percolates down through the snowpack. But his greatest contribution to the mission was his longevity: having been a JIRPer since the 1980s, he knew what the glaciers looked like back then.

Campbell told the group that we'd likely be flying into Camp 18 the next morning. How long we would be there, or how we would get out, was up to the weather. The group clicked into action. Bags were unpacked, repacked, then unpacked again. Drone batteries were charged and vacation alerts turned on. There was no cell signal or internet on the ice. Communication was with a satellite phone, VHF radio, or the mailbag that came with resupply drops every few weeks.

As our heroes readied themselves, I found Wilson stroking his beard, reading a speech given by Maynard Miller back in 1971. It was titled "The Environmental Crisis: What Is the Question?" and was addressed to the Ninth National Youth Science and Humanities Symposium, held at the US Military Academy at West Point, New York. Wilson rested his index finger on a passage and slid the papers to me:

> The environmental crisis is, in effect, the prelude to our "sociological ox-bow" which must be cut off by the mind and heart if man is to achieve a dynamic equilibrium or "steady state" in his society—a state which is not forced upon him by assured pain and suffering if left to nature's inexorable self-regulation. We should not be naive about this. Nature remains oblivious to man; but, to survive, man must not be oblivious to nature.

7

Life and the Living Dead Inside the Glacier

Campbell, having again extended yesterday into today working on a grant, woke me at seven thirty the next morning. "Bus leaves at seven forty-five," he said. Everyone was packed and fed when I stepped into the garage. Eggs had been cooked; dishes washed. Faculty ambled about, heads bowed, staging backpacks and duffels, locking Pelican cases filled with scientific instruments worth tens of thousands of dollars. I was still in my long underwear when Campbell announced, "Van leaves in eight minutes."

To this day, I see those eight minutes as a temporal wormhole. Somehow, I ate breakfast, repacked, double-checked the JIRP checklist, collected electronics from chargers, brushed my teeth, unpacked to find a sweatshirt, packed one final time, and tossed my gear into the back of "Blue Steel," a 1980s kidnapper van with room for fifteen. Dings and scrapes on the body and the musty smell of climbing boots attested to the number of campers who had ridden in it. (The program's other vehicle was a rusted-out Ford Explorer dubbed "The Exploder.")

We passed hundred-foot spruce trees lining the driveway and merged onto Route 7, "Glacier Highway." Roadside backyards were crowded with utility sheds, four-wheelers, pickup trucks, and neat stacks of cordwood. Near the ferry terminal, two dozen bald eagles waded around the mouth of a creek. Fifteen minutes later, we sheepishly stepped onto a scale with our bags at Coastal Helicopters headquarters—to calculate how much excess gear and cellulose we were carrying. A disinterested young woman with curly black hair called out the weights to pilots, who winced every time the number broke three hundred. (Which it definitely did for me and my luggage.) We would not be the only cargo that day. A thousand pounds of frozen hamburger, beans, ketchup, trash bags, batteries, gasoline, and high-tech science gear sat on two pallets near the runway.

The terminus of the Mendenhall Glacier. This flight reminded me of my first helicopter ski trip in Alaska—in Valdez with Doug and Emily Combs. We took off from their home base at the Tsaina Lodge near Thompson Pass and skied a few runs, then flew over an accident site where a snowboarder had fallen six hundred feet from a broken cornice. He died, but his friend who also fell somehow survived. As it turned out, the pilot who flew us to Camp 18 this day had worked with Doug and Emily and flown many of my friends and colleagues to Alaskan peaks.

When we finally got the call to board an hour later, Campbell, Balter, King, and I walked across the tarmac in a very *Right Stuff* procession to an idling Bell A-Star helicopter. The rain had finally lifted, and the morning was warm and clear. The pilot, who I learned later had flown many of my friends and colleagues to Alaskan peaks to ski, told the women to sit up front with him. He nodded at Campbell and me to load into the back. Minutes later, the heli lifted off over the tiny grid pattern of Juneau. Behind us lay cruise ships and tourists; out the windshield was what Mary Shelley dubbed "the everlasting ices of the north" in her cryospheric novel, *Frankenstein*.

"That terminus looks terrible," Balter said, looking down at the Mendenhall Glacier. Cruise-ship-size chunks of ice hung mid-tumble above Mendenhall Lake.

"It's not going to be there for long," Campbell replied.

We flew north past the Mendenhall Towers to the Taku Towers. A few minutes later, we saw the giant expanse of the West Branch of the Taku Glacier and from there the Matthes and Echo Glaciers. Herbert Glacier slid past Triangle Peak in what looked like a frozen waterfall a mile wide. The Gilkey and Bucher Glaciers met in a confluence just like a river with ripples, waves, and eddies. Fifteen minutes into the flight, everything below was brilliant white. The clouds and sky were frozen perfectly still as well. We had left the world of the living and entered another dimension. It was like flying into a postcard—a single, frozen frame from the great flow of history, geology, and ecological morphology.

Campbell pointed to a rocky knob two thousand feet above Gilkey Trench. Tiny ants scurried between buildings as we approached. They were preparing for something, looking for things, stashing things. This was Camp 18, where JIRPers had escaped to—and had been living in the rain for fourteen days, waiting for the skies to clear and helicopters to arrive with much-needed supplies, staff, and their fearless leader. I was afraid we were going to squash some of the campers as we landed, but they were well trained in heli protocol and kept their distance until the rotors stopped spinning. Then it was all high-fives, hellos, where-have-you-beens, and *how was Antarctica?!* Campbell told me that Camp 18 had the best view on the icefield. What lay before us, however, was not a view. It was an experience—of peering around the edge of the world at hundreds of square miles of flowing ice and scrubbed bedrock with hardly any vegetation of any kind. There were no trees, no boulders, no visible life at all, just bedrock and microflora. The glaciers had squeegeed everything clean, leaving behind only granite fractured by what Dennis called the "merciless melt-freeze cycle."

It was hard to believe that some of the most important, time-sensitive questions regarding climate change were being answered here. The nine structures that constitute Camp 18 are not real buildings; they are sketches of buildings—silvery shanties clad with corrugated steel. The shelters were not constructed with any code in mind other than what supplies were available. The builders' vision of what roofing material could be used for—think papier-mâché—was radical. Bent, sliced, and hammered silver sheets capped ridges and gables, encased studs and soffits, framed windows and doors, wrapping structures the way you would wrap a Christmas present.

Ironically, the campus was built by the same brilliant minds currently plotting a course to save planet Earth. Over the years, M3 traded manual labor for the right to study in his 1,500-square-mile laboratory. It was a decent trade. Seventy years later, most of the buildings are still standing. (Except for one that a student named Sam—we will get to him in a bit—nearly tipped over in a frenzy of JIRPness, singing and dancing with fellow—sober—students until the structure bounced off its foundation and lurched downhill.)

A member of the support staff guided first-timers around the campus, pointing out a few 1960s snowcats parked on the edge of a

Ain't No Couplin' Allowed

I don't care how cold you get
I don't care if you're real wet
I don't care if you're doing MET
Ain't no couplin' allowed!

When we move across the ice
You might see someone and start to think twice
And may consider rolling' the dice
But ain't no couplin' allowed!

Travellin' 'round the nunataks
You think of emotions, forget the facts
Don't get excited and lose you hats
Cause ain't no couplin' allowed!

JIRPers ski and JIRPers hike
They're lookin' for friends left and right
Better to love than to fight
But ain't no couplin' allowed!

People say it's no big deal
But as director this problem's real
Can't tolerate no touch and feel
Ain't no couplin' allowed!

All those JIRPers are well and strong
They're working hard all day long
But when they're lonely they sing this song
Ain't no couplin' allowed!

This was posted on the kitchen doorframe at Camp 18, where campers line up in the morning for Spam and eggs. I asked one of the staff if they actually regulated relationships on the ice. She replied, "Yeah, right!"

cliff and outside bunk rooms, outhouses, the kitchen, and library. Sleeping arrangements were bunkbeds. With seventy faculty, students, and support staff in camp that week, all the beds would be full. Mine was directly under Kiya Riverman's. I followed her lead and organized everything I needed to survive on the ice—crampons, headlamp, toothbrush, notebook, headphones for sleeping—then stepped out to find someone who could explain what everyone planned to do for the next week.

The amount of data I had absorbed over the last twenty-four hours was anxiety inducing. Considering I had dropped in essentially midsemester, most people were too focused on their research to answer more of the big, probably naive questions that I had. Like how could these studies and experiments help save winter? Could they stave off catastrophic climate change? If not, what specifically were we trying to do here on the icefield? Besides losing forests and billions of gallons of runoff, how exactly would the end of winter kick off Jon Riedel's metaphorical avalanche?

The ever-welcoming presence of Balter seemed a good place to start. She led me to some flat boulders perched directly above the Gilkey Glacier and used the backdrop as a chalkboard to explain what ice can tell us. Balter had tromped across Svalbard, Antarctica, the Falkland Islands, Scotland, Colorado, Alaska, Greenland, and a dozen other freezing cold places—including Upstate New York—for the last decade. Her aim is to discern why and how ice ages came and went for millions of years and how much ice had been left in Greenland and Antarctica when the climate had warmed in the past. Her latest post was at Columbia University's Lamont-Doherty Earth Observatory, where she was earning her PhD and studying under climate geochemist Joerg Schaefer.

It speaks both to the plenitude of New York City and to the relatively tiny circles that city dwellers like me travel in, that (1) I had no idea that Columbia had a climate and glaciology program, (2) I *teach* there, and (3) Balter and I had never met, though we were likely within a few hundred yards of one another on many occasions. While I navigated potholes on my morning commute, saddled with a backpack of books, Balter rode a shuttle bus to the Lamont campus eighteen miles north of Manhattan. There she spent eight to ten hours in the lab changing out chemicals, crushing rocks that someone gathered in a Frozen Zone thousands of miles away, and dissolving them in hydrofluoric acid. Samples were sent to an accelerator mass spectrometer at the Lawrence Livermore National Laboratory for dating. Combine the dates with carbon content in the atmosphere and how much water on the planet was locked up in snow and ice—the last two sums derived from ice core analysis—

and you get a fairly accurate picture of the climate, sea level, and even ice extent of yore. Which is exactly what Balter was after in 2017 when she traveled to the East Antarctic Ice Sheet.

Her mission was to study the history of ice in the central Transantarctic Mountains. The scale she was looking at was not thousands but millions of years, using a dating technique that measures cosmogenic nuclides created by high-energy particles. These cosmic rays enter earth's atmosphere from outer space. When they hit a rock, they induce tiny nuclear reactions in minerals within the rock, creating isotopes* that you wouldn't otherwise find on earth. The most common one Balter finds is beryllium-10, which has become a kind of stopwatch indicating how long a rock has been exposed to the sun. For example, if a rock is covered by ice, few nuclides are formed. When the ice melts away and exposes the rock to the sun, more nuclides are formed, giving us a timeline of where the edge of ice sheets and glaciers were located throughout history.

"The most common one that we think of is delta-18-O," she said. She was talking about $\delta^{18}O$, or the change in the ratio of "heavy" oxygen (^{18}O) to "lighter" oxygen (^{16}O): "When you measure that in forams, which are little critters living in the ocean that die and become part of the sedimentary record, you can get an idea of the temperature of the whole ocean in the past and how much ice volume there was on earth. The sixteen-oxygen is the lighter one, and eighteen-oxygen is the heavier one. When you have more ice on earth, more ^{16}O is locked up in it, so the relative portion of ^{18}O in the ocean goes up."

The process of collecting this data is tedious and slow, with thousands of measurements eventually adding up to one story—a kind of sorcerous text message from the past, packed with information

* Isotopes are defined as atoms of a chemical element with the same atomic number and similar chemical behavior but a different number of neutrons in their nuclei.

on how heat and cold dissipate around the planet, where snow fell, how oceans warmed, and how carbon dioxide affected it all. For example, one new theory arising from isotope dating rejects the idea that the sun or oceans or even, directly, the atmosphere kicked off the succession of ice ages. Rather, rocks themselves, unearthed by plate tectonics, sent the planet into a sudden cooling trend. Around fifty million years ago, the Indian and Eurasian plates collided and formed the Himalayas. Newly exposed rock absorbs carbon dioxide, and with so much rock thrust up from the depths, the cooling effect could have been drastic—forming ice on the poles and glaciers across the Northern Hemisphere, which, in turn, reflected more of the sun's heat and cooled the planet further, ultimately initiating the Ice Age pattern and even winter as we know it.*

Balter did not find answers to all my questions on her 2017 Antarctic expedition, but she did unearth something the cryo-community has been grappling with since. "I was looking basically at the middle of Antarctica, which is so cold and so dry," she said. "It's a place you wouldn't really expect to see a ton of change but might capture the health of the whole ice sheet. So, maybe the margins have retreated a little bit, but if the ice sheet were to have collapsed, we would see it there and if not, we wouldn't. So I found that during the Pliocene—and even the Miocene, five to twenty million years ago—when the temperature was even warmer and carbon dioxide was thought to be even higher,... during both those times, ice, at least in that location that I was studying, was a similar size as it is today, or a little bit larger."

Her observation would mean that during some of the hottest, most carbon-rich atmospheric conditions ever seen on earth—when

* A slowdown in plate tectonics in the late Miocene, fifteen million years ago, may have cooled the planet again by decreasing the release of carbon dioxide from volcanic activity on the ocean floor, cooling the planet by fifty degrees Fahrenheit.

giant camels romped through swamps in the Arctic Circle and scientists of today assumed that there was not a shard of ice on the planet—there were still glaciers on Antarctica. And they might have been *growing.* Which flies in the face of pretty much every climate change projection I had ever read. It also brought into question the very meaning of glacial data that had been retrieved for the last century—and whether we were misinterpreting M3's tea leaves.

Wilson Clayton's Spiritual 401(k)

My discussion with Balter left me with even more questions, but my hunt for answers would have to wait. All new arrivals were summoned to a rock amphitheater near the dining hall for a safety course. The course is mandatory because your chances of survival on the glacier without it are surprisingly low, especially that year, with new crevasses opening up. Glaciers flow like water, more like brittle water that snaps and breaks and melds back together under pressure, creating unstable bergschrunds, ice shelves, seracs, and, most importantly, crevasses. Snow typically covers up cracks like spackle on drywall—just thick enough to conceal them, but too thin to support a human step.

"This is what we will try not to do," JIRP's lead mountain guide, Ibai Rico, said to the crowd. The setting could be described like this: a group of interesting, slightly dorky scientists and climbers sitting before an alpine aerobics instructor. "Extremely European" would be another way to report Rico's appearance. He was an endlessly talented Basque glaciologist, mountain guide, professor, and amateur standup comic. The thing I couldn't get out of my head, though, was how much he looked like Adam Sandler playing the role of a swashbuckling Euro glaciologist-professor-mountain guide—complete with dark tan, heavy accent, headband, white sunglasses, Hawaiian shirt over burnt-orange polo, and perfectly copped five o'clock shadow to accentuate his chiseled Basque looks. He was an activity director leading a group exercise class on a cruise ship. He was an ayahuasca guide at a Guatemalan treehouse retreat. Our visualization today: make all our friends not die.

Rico's first lesson was to show us how to walk in a line roped together—so if one person falls into a crevasse, everyone else on the rope can stop the fall. The basic idea is that to save themselves, the group members will dig their toes, axes, and fingernails into the snow, by proxy saving the victim. The next trick was learning how to tie a bowline and a Prusik knot, again, in a manner that does not kill your friend tied to the other end. Rico workshopped comedic bits on the crowd as he taught. After each lesson, to punctuate what would happen if you messed up, he would spread his arms and make the faint sound of a person falling to their death. "You don't go around twice the tail of the rope," he said, "and . . . *aaaaaaaahhhhhhhhhh.*" (Pause for laughter.)

Halfway through the class, he came up with another bit—an acronym he made up for faculty members who, like Campbell, had enough climbing experience to make the rank of Faculty Accredited Trainer (FAT). Rico then told the crowd three or four times that FAT people could be trusted—pause—then seemed confused when the laughs didn't roll in. In the end, he pulled off the class, and we learned to walk safely across the icefield. Throughout the entire course, terrific groans and rumbles from surrounding glaciers reminded us how dynamic the ice was and that it had been flowing for millions of years, pushing toward the equator like an icy windshield wiper, then retracting again, a proscenium of greenery cropping up in its wake.

The ice was moving faster than ever now. On the flight to Camp 18, the pilot mentioned how brown the trees looked and how pockmarked the glaciers were. In just the last sixty years, the nearby Llewellyn Glacier had retreated more than a mile and a half. The

Norris Glacier had shrunk by a mile, and the Gilkey, right below us, had retreated more than two miles. Even the mighty Taku Glacier, once the last advancing glacier on the Juneau Icefield, was in full retreat.*

Students began daily chores after the class: cleaning outhouses, arranging blue tarps to melt snow for drinking water, cleaning the kitchen, burning trash in a very nonsustainable toxic inferno. (Campbell was writing a grant to make the camps carbon-neutral.) I headed to the bunkhouse to grab my notebook and found Clayton standing in the middle of the room holding a broom. He was not going to answer my big questions. I wasn't sure who would; I just knew I needed someone to distill the mass of information I was about to bring home.

"It's always disheartening sweeping these carpets," he said. To demonstrate this level of disheartenment, he ran the broom over the rug. A cloud of dust rose into the air, then settled back down on the rug. It would not be too inaccurate or unfair to suggest here a metaphor for the golden years of Wilson Clayton's life. Which is not to say that the man was spinning his wheels. Or maybe he was, but he was doing it by choice. "This is kind of my retirement plan," he said. "JIRP in the summer. Ski patrol in Colorado in the winter."

Clayton started to delve into how a degree in geology led to a successful career in the "booming groundwater remediation scene in the 1980s." He was only two paragraphs deep when a frantic student interrupted his story. She was looking for trash bags. Not for garbage, she explained. It had just been announced that Fancy Dinner, the nexus of JIRPness, would take place that night. After weeks

* The icefield was also the subject of John Muir's musings. "Over the mountains and over the broad white bosoms of the glaciers the sunbeams poured," Muir wrote in his 1915 collection of essays, *Travels in Alaska*. "...Rosy as ever fell on fields of ripening wheat, drenching the forests and kindling the glassy waters and icebergs into a perfect blaze of colored light."

of rain, the sudden blue skies, sunshine, and a thousand pounds of provisions inspired the feast. Right at that moment, a hundred frozen hamburgers were thawing on a charcoal grill.

By the time I got back to the mess hall, all scientific programs and processes had been suspended. Former student and current mascot and bon vivant staffer, Max Bond, was shirtless, flipping burgers on the grill. This was the same Bond who had knocked the dorm off its foundation, himself a pillar of JIRPness overflowing with energy, smiles, curiosity, jokes, songs, and innocent mischief. Other students in the mess hall cut trash bags into gowns, neckties, blazers, and blouses and dug through two boxes of dress-up clothes kept in the back of Camp 18 for just such occasions. Rico arrived, wearing an alarmingly tight pair of bright red pants and a cutoff jean jacket—now a dead ringer for Bruce Springsteen as ski instructor. When a young student showed up wearing a perfectly tailored duct tape bow tie and collar, Rico leaned over to me and whispered, "What a bunch of beautiful nerds!"

Two support staff carried four long tables and a half dozen benches from the dining hall and set them up on the granite helipad. As the sun set and yellow light poured down Bucher Glacier, the ice turned periwinkle blue, then yellow, then red. The silvery campus faded into the dark, and a nest of light pooled behind the rockbound summit of the Tusk. All fifty campers and faculty sat at the thirty-foot-long banquet table, eating burgers and salad and talking about their lives, their partners, their projects, and the grandeur of Camp 18. The scene reminded me of the dance that Gleason, Swensen, and Krause had performed on the flanks of Mount Adams. Right then it seemed that everyone at the table was exactly where they should be and that their presence was undoubtedly a good thing for humanity.

Kate Bollen, our camp manager for the week, had organized an art show after dinner—a show that most people reacted to with

A bunch of beautiful nerds at "Fancy Dinner."

varying levels of dread. But like most events at JIRP, the real thing was quite a surprise. The first exhibit was a performance art piece, in which a very young undergrad woman from UC–Berkeley lit into a brilliantly filthy rap, stringing together inside camp jokes. Bond morphed into a human beat box, and the girl grabbed a VHF radio to distort her voice for the hook as the entire shack bumped up and down. Bond took the mic next, singing an Arlo Guthrie–inspired ode to the Hilton Hop, the building that he bounced down the hill. After a brief introduction to some excellent sketches that students had drawn in their notebooks—all leaned up against the mess hall windows—most of the faculty headed off to bed.

One of the many rites of passage on this geological tour was, on rare clear nights, to sleep on the rocks under the stars. That night offered the first clear sky in weeks, and one by one, students and teachers dragged their sleeping pads and bags out onto the ridge for celestial communion. I imagined what the experience must be like for them, with their understanding of the earth's crust, ice, and invisible particles spinning off the very stars overhead. What an acid trip of awareness that must have been, gazing up from their world into a 180-degree ocean of beryllium-10, oxygen-18, solar systems, black holes, icy comets, and flaming suns. How disorienting it must be, I thought, for these young scientists to live in two worlds simultaneously—one foot set in human time as they navigate traffic and relationships and grant proposals, the other probing the dark morass of deep time like a child feeling for the deep end of a swimming pool.

Clayton was on his way out when I walked into the bunk room. He didn't say a word. This was his moment, the spiritual 401(k) that he planned to ride to his final day. I watched him disappear over the rocks, weaving between little lumps of sleeping-bagged students, out past the farthest ridge to a sloping granite cradle, where he lay down beneath the bluish light of the universe.

"The Area Is So Dead Even the Devil Has Left!"

The wind rattled the corrugated roofing so violently the next morning that I thought the glacier was calving. I ran outside to see what was happening, but it was just wind from another storm front. A line of mare's tails stretched across the stratosphere. It was so clear that I could see Mount Fairweather rising out of Glacier Bay National Park a hundred miles west, beyond the glacially scoured Favorite Channel. It was five in the morning and looked as if the sun had been up for hours.

Students who had slept outside stirred in their sleeping bags and began the chilly walk back to bed. I managed to go back to sleep for a couple of hours until a student on wake-up duty walked in blasting the Foundations' "Build Me Up, Buttercup" on a portable speaker. I put on my long underwear, mid-layers, wool socks, ski pants, down jacket, waterproof gloves, sunblock, and glacier glasses. The dining hall was half full when I got there. On the menu that day were fried Spam, pancakes (one each), and a single spoonful of hash browns. I was lucky enough to sit across from Christine Foreman, a chemical and biological engineering professor from Montana State University. Dr. Foreman was at JIRP to conduct research on yet another extraordinary quality of ice—its ability to harbor life. Scientists have long studied the subnivium, a stratum of life that exists within and below the snowpack. The subnivium enjoys a microclimate created by heat and moisture escaping the ground. Snow crystals elongate into hoarfrost as moisture evaporates and rises, forming a cozy pocket near the bottom for creatures like ruffed grouse, mice, porcupines, voles, and an array of fungi and mosses. Snow is an excellent insulator and keeps the temperature of the subnivium around thirty-two degrees Fahrenheit—often thirty degrees or more warmer than the air above.

Microbes in ice are different, Foreman said. Some microbes in Antarctica use a giant protein as a kind of antifreeze grappling hook to attach to ice and start a colony. Others survive in a nanometer-thin layer of water that forms between ice and a mineral surface. Still others, like the internet sensation "water bears" (tardigrades), live in cryoconite melt holes layered with primordial sand, micrometeorites, and cosmic spherules from interstellar dust, asteroids, and comets.* These water bears and cosmic melt holes are the subject of the study of *panspermia*—or how life may have been transported around the universe frozen in icy comets. Recent research has dispelled the theory that repeated collisions with comets and meteorites provided earth with all its water. It turns out, the material that

* A fun experiment is to collect your own cosmic dust by leaving a magnet in the bottom of a cookie sheet filled with water outdoors during a meteor shower. The micrometeorites will gather around the magnet.

made the planet billions of years ago contained enough hydrogen to create several oceans. But the comet and asteroid hits could very well have delivered frozen packages of life-forms. Samples taken from the 67P/Churyumov-Gerasimenko comet by the European Space Agency's Rosetta space probe included many of the compounds needed for abiogenetic life.

Foreman was at JIRP to study microbes living in micron-diameter veins in ice. They have been found in virtually every ancient ice core analyzed. The life-forms are an organic postcard from the ancient past—organisms once thought extinct but living, quite happily, awaiting the day when warmer temperatures release them. Some live in a state of suspended animation, meaning they remain dormant but viable, given the right environmental conditions. Others exchange DNA and even evolve to take on new traits while they hibernate. Still others, like *Bacillus anthracis,* the bacterium that causes anthrax, could spread ancient diseases around the planet as they thaw. Remarkably, anthrax did spread this way in 2016, when a frozen deer carcass in the Siberian permafrost that was infected with anthrax thawed and killed thousands of deer and one child. In all, it is estimated that the mass of microbes living in the ice right now outweighs the human population by a factor of a thousand.

Foreman stood out the day before as perhaps the only guest at the camp not having the time of her life. Her frustration might have come from the fact that her collaborator and spectrometer operator canceled at the last minute, leaving her and her assistant, Markus, literally out in the cold. Markus was an Austrian wild man whose very Germanic and abrasive sense of humor—which I came to appreciate dearly—seemed to offend nearly everyone in camp and might have been another source of Foreman's frustration. But the two were married for the week, to each other and to a very complicated and expensive impedance spectroscopy sensor packed away, along with two coolers to store ice cores.

Their research was also funded by NASA, not so much because the administration is committed to saving the world's glaciers but because most of the planets that humans are casually considering absconding to are frozen. And on those frozen planets there is ice, in which, Foreman told me—cheeks flushed, eyes wide—there might be life. Life similar to the kind living right there on the Gilkey Glacier.

"There is a region in Antarctica that they call the Dry Valleys," Foreman said.

"One explorer wrote in his journal," Markus interrupted, with his heavy German accent, " 'The area is so dead even the devil has left!' "

"But they found life in the ice," Foreman continued. "And now the valley is one of the oldest ecological preserves in Antarctica."

Foreman has visited Antarctica several times and has become an Antarctica-phile. "They preserved the sheds down there that Shackleton and other pioneers and explorers lived in," she said. "There's rotting seal blubber still there and old scientific tools. The cots are tiny. They still have rotting reindeer-hide sleeping bags. They found a stash of Scotch beneath the building. They put a probe into it and sent it back to the lab to figure out where it was from and how it was made. Then they made another batch to commemorate the explorers."

Foreman's research has gotten the attention of some of the biggest geological institutions. She sits on a panel that will decide what future NASA missions in space will look for. She was most interested in equipping upcoming NASA missions to Mars with ice-core drills to search for life in the ice or in oceans beneath the surface of the planet.*

---

* NASA plans to send multiple drilling missions farther into the solar system. The missions include a proposal to drill through more than eighteen miles of ice on Europa, one of Jupiter's moons, to search for signs of life. A proposal to test a prototype drill on the Juneau Icefield is also pending.

8

The Enchanted Divide

Campbell interrupted my exploration into frozen alien life to tell me that we would be skipping the second day of safety training to visit another satellite research site. To glaciologists, the icefield divide serves as the provenance of the glacier itself. The divide is where snow first accumulated, compressed, and began to spread. It is also the place where radio glaciology specialists like Campbell can find straight, orderly ice layers that have not been deformed by glacial movement. The snow and ice layers here look like a fresh stack of pancakes. Miles away on a moving glacier, the layers look like a stack of pancakes after someone who likes pancakes very much has soaked them in syrup and cut them up. The difference is so stark, and the results from drilling at the divide so superior, that most of Campbell's work now revolves around using impulse radar to find divides for drilling operations.

He had picked drill sites in Denali National Park and the Canadian Yukon and recently has been working in Antarctica with a hundred scientists on a $60 million collaboration between the United States and the United Kingdom. The mission is to determine the stability of Thwaites Glacier—the "Doomsday Glacier"—which scientists say props up and drains much of the West Antarctic Ice Sheet. Because the ice sheet is marine based—set on bedrock below sea level—you get *reverse bed-slip* if the ice shelf that supports up-glacier ice breaks apart. In other words, reverse bed-slip occurs if the glacier loses its grip on the rock in front of it and becomes a floating chunk of ice that falls into the sea. (Sea levels could rise up to ten feet if Thwaites fully deteriorates.) With warmer ocean water now melting Thwaites, the glacier's velocity into the sea has increased 50 percent between 2010 and 2021. "You've seen stories about ice chunks the size of Rhode Island breaking off," he said. "That's potentially Thwaites Glacier, and everything behind it that's flowing down."

In the last few years, Campbell has been finding drill sites that pass completely through a glacier and pick up bedrock samples from beneath it. These rock samples can be tested to calculate the last time the glacier receded. "We're calling it the 'dipstick' technique, a term coined by Dr. [Harold] 'Hal' Borns from the University of Maine forty years ago," he said. "We're spot-checking different locations to say, 'Okay, we think the ice elevation should have been at this point ten thousand years ago. Let's drill a hole and collect a rock sample and see if that matches.' " Other crews from the University of Maine simultaneously collected fossilized animal bones on previous shorelines and determined rates of shoreline change due to uplift of the crust—as the ice sheet retreats—and sea level changes, estimating where the coast of Antarctica was during various epochs.*

Campbell loaded several black Pelican cases into a giant blue sled hitched to a snowmobile as his student assistant for the day, Eva, and I clicked into our skis and skinned to the top of a ridgeline behind the camp. Or I should say that Eva skinned and I slogged many yards behind with my gimp knee. Campbell took pity on me at the top, strapped my skis to the front of the snowmobile, and offered me a seat on the sled while Eva grabbed a towline and dragged behind for

* Bedrock uplift might counter a Thwaites Glacier collapse if uplift rates (which keep bedrock in contact with the ice, thereby maintaining more friction to slow down ice flow) keep up with glacier retreat.

forty minutes. The route passed through a white wasteland cleaved by a dainty snowmobile trail marked by flags the divide team had previously installed. The small piles of rock we drove past were once summits of the Storm Range, which divides the United States and Canada. Covered by more than three thousand feet of ice, they were mere hillocks now. A yellow sign in the middle of the field pointed to Camps 8, 9, 18, and 26. We took a left toward Camp 26 and followed the track to a cluster of tents on the divide.

The camp looked like a moon base as we approached—including a few astronauts wandering around in puffy snowsuits. The surface of the divide was flat and pocked with volleyball-sized melt holes called *sun cups*. Six tents stood in a line, with a large cook tent and something that looked like a teepee on the far end. A red pole next to the teepee was the thermal ice-core drill that we had come to see. Inside was a twentysomething, goateed man named Grant Boeckmann, who slowly lowered the apparatus while watching several digital displays. The control box did not look like something NASA made. It looked more like a gadget a hobbyist engineered at Radio Shack. Boeckmann adjusted the tension of the cable—trying to keep it taut as it lowered the drill head—by turning a stainless steel disc. His partner, Jim Koehler, monitored tension on the line, walking into and out of the tent to make sure the cable fed out properly.

There is nothing exciting about drilling, Boeckmann said, except for the results. It can take a few minutes to drill down one inch, and the process is outrageously precarious. Ice density is not consistent, even within the four-inch diameter of the hole. Holes that are miles deep will veer off drastically if the bit tilts a fraction of a centimeter in any direction, usually causing it to get stuck or break—and ending the mission. Boeckmann said that a major operation in Antarctica took a year to set up and another year to drill down a mile. But when the bit broke recently, the equipment was rendered useless and the project had to be scrapped.

If the divide were a small town on the Western Slope of the Rocky Mountains, this drill tent would be the 7-Eleven where everyone hangs out. In this fantasy world, Grant Boeckmann (pictured here) is the high school quarterback who graduated two years ago but still hangs out at the "Sev" because it's his turf and he knows the place better than anyone else does—and what the hell else is he going to do in this small town?

For the next hour, Koehler told me how boring the job was; meanwhile, Boeckmann kept turning the disc. They had been drilling for over a week and were about three hundred feet down. Every fifteen minutes or so, Boeckmann brought the head back up with a nine-foot cylinder of ice that two students transported to a cradle and inspected for ash layers and air bubbles. At three hundred feet they were seeing ice that fell as snow sometime around the War of 1812. Six hundred feet down, they'd be inspecting ice formed by some of the same snowstorms that buried colonial outposts at Jamestown and Providence.

While these samples provided a picture of climates of the past and terrific evidence of human history and evolution, the climate community was after something else: a weather report. Global climate models require vast amounts of data collected from all over the world. They also need data from the past—the ancient past—so they can input known values and use hindcasting to hone accuracy. This data comes from ice cores, rock samples, and fossils like the microarthropod Collembola, which thrived in Antarctica when it was ice-free and which declined during ice ages. Input all this information, and scientists can test their model by hindcasting, say, March 4, 1964, and then compare the result to actual records. Using ice-core data hundreds of thousands of years old, they can estimate climate throughout the eons.

The European Union recently launched the largest, most detailed climate model ever made. Destination Earth will be a "digital twin" of earth's atmosphere, ocean, ice, and land with pixels a half mile across—many times more detailed than any existing model. Using data collected at JIRP and beyond, designers hope the model will be able to predict droughts, floods, and even fires years before they take place. It will also track the human contribution to climate change—atmospheric pollution, crop growth, energy production—comparing climate policies from different nations and offering citizens and governments a clear mandate on what needs to be done to avoid the worst outcomes of global warming.*

The drilling operation at the divide was a test for a planned expe-

---

* Destination Earth (DestinE) will take ten years to build and will use three *exascale* computers spread across Europe, each computer capable of making a quintillion calculations per second. The result will be a model that can render basic climate building blocks like the formation of clouds and swirling ocean eddies, which transport heat and carbon. Monitoring things like energy use, traffic patterns, and human movement (using cell phone data), DestinE should be able to determine, for example, how cuts in ethanol subsidies in Brazil might limit deforestation in the Amazon.

dition in Antarctica. Cores from active research projects typically end up in the National Science Foundation (NSF) Ice Core Facility in Boulder, Colorado—a 55,000-cubic-foot freezer holding 56,000 feet of ice cores at minus thirty-two degrees Fahrenheit. There, technicians use a heat knife to split ice into ten specimen tubes. Sections with air bubbles are crushed in a vacuum tube so that the gases can be sucked out and analyzed.

"It's a pretty menial job," Koehler said for the third time. Koehler was the veteran of the crew. He was middle-aged with a salt-and-pepper crewcut and a thick scruff. It looked as though he had been wearing his polyester crew top, tan snow pants, and hiking boots for at least a week. Unlike the rest of the group, he wore no hat or gloves outside, and he had the build of an oilfield roughneck. A profession that is not so dissimilar, he said.

Koehler was from Wisconsin and had no problem talking to, and sharing sensitive information with, a complete stranger. Like how Russian core drillers in Antarctica use vodka to lubricate drill holes. Or how US teams dump barrels of black-market Freon into holes to keep them from freezing shut. Koehler ran his drilling company like a lawn-mowing business. "You gotta start an LLC," he said apropos of nothing. "I got my own company; I go where they tell me. I've been in Antarctica seven winters in a row now and am contracted for the next two."

A professor in college got Koehler into drilling. His first gig in Antarctica was fifteen years ago with the IceCube project. The project is based on one of those concepts that scientists breeze over but that make you realize (1) how little you know about the planet we live on, and (2) the ludicrous reality of living on a hunk of iron caught in an air bubble surrounded by the vacuous, black, infinite nothingness of space. For example, did you know that some of the tiny particles from outer space flying through our atmosphere and planet and us every second of every day are neutrinos? Did you

Boeckmann did not build this contraption with parts from Radio Shack. It's a real ice-core drill-control unit, capable of fetching ancient ice a thousand feet deep, which it did a few weeks after this photo was taken.

know that these neutrinos are so small that they literally fly between atoms in our bodies and the planet without hitting anything? And that when, once in a blue moon, they do hit something, the collision makes a little flash of light, like a microscopic flashbulb going off?

Did you know that there is an ongoing experiment in Antarctica in which eighteen hundred glass balls—think Japanese fishing buoys—were hung in thirty-six identical, two-kilometer-deep, three-foot-wide holes in the ice? And that these holes were fitted with photo sensors to capture the off chance that three neutrino collisions lit up simultaneously—so that scientists could triangulate the origin of the particles? Like a hunter finding his way with a compass fix on three mountain peaks? Did you know that Koehler drilled these holes? And that, in September 2017, after thirteen years of silence and many millions of dollars, the trick actually worked?! IceCube tracked a high-energy neutrino back to a blazar—an active galactic nucleus—known as TXS 0506+056, located 5.7 billion light-years away in the direction of the constellation Orion. "They got more funding after that," Koehler said dryly. "They should be good to go for another ten years."

Self-Assembly and Kepler's "Living Soul"

The crew at the divide ambled around the camp most of the afternoon, gathering samples, entering data, and making coffee with a secret AeroPress hidden in the mess tent. The drill sank lower into the divide, painfully slow, melting through decades. Compared with the poles, Alaska was so warm right then that reading ice layers was getting difficult. Meltwater had seeped down through the glacier, erasing chemical signatures.

Around two in the afternoon, a bolt fell out of the thermal bit and got stuck at the bottom of the hole. The bolt was, of course, lying precisely across the melt ring, slowing drilling significantly. The operation stopped. The project-ender the crew had been testing for had happened, and now they began—quite excitedly—to do what they were actually paid for: save the experiment by MacGyvering a solution.*

As the group laid out different strategies to retrieve the bolt—one idea used a basket on a tether with a GoPro camera attached, another used the retractable teeth on the drill itself—Brad Markle assembled a crew to gather surface samples from the divide. He

* The crew did continue to drill and eventually made it just shy of three hundred meters, providing Dr. Foreman with a marvelously unique sample to fit into her cooler and study at home.

invited me along, so I clicked into my skis and followed them. Going anywhere on the divide is comical; the ice cap is a flat white disk with few landmarks and seemingly no end. Markle pointed at a low hill that could have been fifty yards or fifty miles away. After skinning for twenty minutes, we appeared no closer to it and no farther from camp. The job three students undertook seemed equally futile. Every fifteen hundred feet, one of them knelt down with a dinner spoon, scooped up a snow sample, and placed it in a sample jar, on which another student marked the time, date, and GPS coordinates. Twenty minutes later, they did it again. A teaspoonful of snow on what seemed an infinite stratum of ice did not seem like it could tell us anything about, well, anything. But after seventy years of tiny spoonfuls scooped by M3's students, all placed into perfectly labeled jars, the wealth of data JIRP has gathered has made it one of the most important glacial research resources in the world.

Markle gets this. He watched every vial, every spoonful, and corrected the slightest mistake. It was as if each sample held the answer to a decades-long theory that he had been trying to prove, which it kind of did. But to keep up that kind of concentration ten hours a day, for six weeks at a time, living in a tent in the rain and eating Spam and eggs for breakfast, well, it was impressive.

We continued our hunt along the divide, step by step, spoonful by spoonful. Beneath our feet, snow compacted, melted, morphed, and bonded, beginning its long journey down and outward in a glacial branch. My mind wandered in the milky white snowscape, and I thought about Johannes Kepler in the 1600s and his consternation at what caused ice crystals to form patterns the way that plants do. Like an inanimate object mimicking life. *(Heresy!)* He discovered properties of other inert things, such as how planets orbit the sun and what propels ocean tides. But snowflakes stumped him, and he ultimately gave up: "The shapes of snowflakes are by no means to be deduced from the operation of soul in the same way as of plants."* The Roman naturalist, historian, and soldier Pliny the Elder—who first applied the Greek word for crystal, or *krýstallos,* to quartz—had a difficult time explaining snowflakes as well. He claimed that quartz was a form of ice so frozen it could not melt.† He was wrong about that, but there is a similarity between this mineral and ice. Quartz, ice, copper, and diamonds are all crystals, defined as a substance in which atoms or molecules are arranged in a repeating internal pattern. It is this arrangement, this so-called self-assembly, that had never been fully explained. What causes a thing to put itself together? Either a snowflake or a flower? What is the guiding force that shapes a crystal of sugar or salt or even the shape of aluminum foil? How about you and me? We are also self-assembled, according to a defined order mapped in our genes. Did Kepler's "living soul" guide our creation? What, then, guides the infinite options a snowflake can take?

I would not get a chance to ask Markle these metaphysical questions. As it turned out, my expedition into the enchanted divide would be my last at JIRP. A broken transmission erupted from Markle's radio about an hour into our walk: a storm was coming; Coastal Helicopters had canceled my flight back later that week. If I wanted to get home this month, I had to be at Camp 18 in an hour to catch the last flight out.

What followed was a comedy of errors, mostly perpetrated by me. It started with a chaotic free-heel ski, which is tricky for anyone but downright impossible with one operational knee. I wiped out twice but managed to get back to camp. The divide camp manager,

---

* See Kepler's 1611 "The Six-Cornered Snowflake," in which he compares the building of a snowflake to the way sailors stack cannonballs on a frigate.

† Like the inert "ice-nine" from Kurt Vonnegut Jr.'s *Cat's Cradle*. Vonnegut borrowed the idea from his older brother, Bernard Vonnegut, an MIT chemist who discovered how to seed clouds with snow, a practice that ski resorts in the US West continue to this day.

Annika Ord, had gotten the transmission and was waiting with a snowmobile. She strapped on my skis and gear, handed me her radio, and took off across the sun-cupped divide so fast that the radio, my bag, and (almost) yours truly bounced off the seat and onto the track a mile out—making it impossible for us to (1) continue or (2) call for help until we found the radio, which we could not. When we finally did and called Campbell and Eva on the Matthes Glacier, Campbell drove over and turned off his machine so that we could make a plan. The plan did not account for his machine's failure to start back up. So Campbell and I loaded onto the other snowmobile, with Eva dutifully in tow again, and left Annika to walk a chilly mile back to camp.

Halfway to Camp 18, we spotted two groups skiing down a distant peak. They had been taking samples and measurements on the summit, which had been climbed only one other time, in the 1960s. The group left long, arcing S's on the yellowish corn snow as they skied one by one. A field of crumbling gray peaks wrapped around them, divided by glaciers and deep valleys. Eva looked on longingly, changing her grip on the rope every few minutes and then finally dropping it at the top of the last ridge. I got off the sled too and skied the final slope to camp with her. It was the second time I'd skied since my surgery. My knee didn't hurt; the nerves were still so damaged that it was mostly numb. After sliding the first section, I angled the edges and leaned into a few turns. The run was gentle, but the surge of gravity pulling me downhill—imagine a swing at the bottom of its arc—felt incredible. I made a few more turns, then let the skis run down the final pitch, stopping at the helipad, where I dumped my gear. It had been fifty minutes since we got the call. I raced to my bunkhouse and packed, made a few quick goodbyes, and scrambled to the dining hall to wait for my ride.

9

The 10,000-Year Window

The helicopter did not come in ten minutes; it was more like two hours. JIRPers milled around the dining room, killing time between classes and chores while others arrived on skis from field sites, awkwardly scratching their way down the ridge behind the dining hall. As I wandered over to one of the promontories to see the glaciers one last time, I recalled something Markle had said. It was before our excursion, before the thermal drill got jammed, a quick moment just after lunch. When I'd asked if we could chat, he ushered me a few hundred yards away from the tents. At a university, this would have been his office, and he would have closed the door and sat behind a desk. On the white thimble of the divide, it was just a patch of snow separating us from the students and colleagues he'd been living hand in glove with for three weeks. The spot was framed by two small crevasses, one of which I nearly stepped into before Markle gracefully offered a hand and suggested that I jump.

This is the kind of person Markle is. Supremely capable, confident, yet understated and compassionate. He was reminiscent, both physically and spiritually, of Indiana Jones* —add beard, long dark

* Apologies for the multiple Harrison Ford references, but the actor has really cornered the market on the adventure-science, sci-fi stud thing.

"The abrupt sides of vast mountains were before me; the icy wall of the glacier overhung me; a few shattered pines were scattered around; and the solemn silence of this glorious presence-chamber of imperial Nature was broken only by the brawling waves or the fall of some vast fragment, the thunder sound of the avalanche or the cracking, reverberated along the mountains, of the accumulated ice, which, through the silent working of immutable laws, was ever and anon rent and torn, as if it had been but a plaything in their hands."—Mary Shelley, Frankenstein

hair, white-framed glacier glasses, and royal blue down jacket. His JIRPness was off the charts. Markle first came to the icefield as a student in 2007, the same year that Campbell arrived. This was Markle's eighth season, and he was nearly finished with his PhD. Titles of two recent papers give you an idea how he spends his time: "Variable Relationship Between Accumulation and Temperature in West Antarctica for the Past 31,000 Years" and "The Spatial Extent and Dynamics of the Antarctic Cold Reversal."

Markle's specialty, paleoclimatology and climate dynamics, often puts him in a tent with drilling crews. "The answers that ice cores can give us are growing," he said. He spends much of his time analyzing the chemistry of ancient precipitation, searching for some of the same isotopes Allie Balter collected. Markle also looks for layers of frozen dust in ice cores—dust that blew off South America and onto the Antarctic Ice Sheet a half million years ago—to help determine climate, rainfall, wind strength, and wind direction at the time. Other signs he looks for include salt that blew off the sea, to deduce whether the Antarctic Ocean was liquid or iced over during any given period.

Like Balter, Markle has been studying anomalies, including rapid temperature changes, in the Antarctic climate. People long thought that the planet's natural climate cycles occurred slowly and in synchronicity. But evidence in Antarctic ice cores now suggests that abrupt changes in the Northern Hemisphere affect the south in immediate and sometimes surprising ways, perhaps explaining the anomalous glacier growth that Balter discovered in the Transantarctic Mountains.* "Climate changes in the tropics can affect the poles in different ways and make them feel like very different places," he said. "Atmospheric circulation has these big cells in the tropics, in the midlatitudes and the high latitudes. And they're all connected. So when something happens to shift the temperature in the Northern Hemisphere, it affects where the winds want to move around, and where they want to circulate energy in the tropics, and then that affects where the winds in these high southern latitudes near Antarctica shift around as well. Standing here, it doesn't feel

* A recent study by Balter's alma matter, Columbia University's Lamont-Doherty Earth Observatory, found evidence that brief epochs like the Little Ice Age and the Medieval Warm Period were not global events and were not felt beyond certain regions.

like what's happening in Hawaii right now really matters much. But yeah, over big enough time scales, it does. The whole atmospheric circulation shifts."

Evidence can be seen during the last Ice Age, when the surface temperature of Greenland spiked ten to fifteen degrees Celsius—in a single decade, several times. There were no man-made greenhouse gas emissions then. There was no significant volcanic activity. Scientists are still unsure of what happened, but they know that every time it did, winds blowing around Antarctica shifted several degrees north. "One of the most surprising things, and a result of all the work being done there, is that it doesn't really seem like anything external caused the Greenland spike. It was just internal," he said. "The climate is complicated enough on its own that it can result in this kind of variability just through different parts of the system interacting with each other."

Climate deniers love to single out tidbits like this: natural causes, not humans, sometimes change earth's climate. This observation is accurate in some cases. Look at the ten-degree temperature spikes in Greenland that Markle mentioned. Or consider thermohaline circulation in the oceans. A branch of the current in the Atlantic—the Atlantic meridional overturning circulation (called AMOC)—pumps warm water north from the tropics to the coast of Europe, keeping the continent artificially warm. Research suggests that a "cold blob" of fresh water rushing from Greenland's melting glaciers has slowed the current and might even shut it down someday—causing a drop of up to twenty-seven degrees Fahrenheit in just two or three decades in Europe, plunging the region into an ice age for centuries.

This kind of natural incongruity—*pattern making*, a physicist would say—is how our very complex climate system works. Unstable systems shape our climate and even our world the same way that instability forms a snowflake. Take everyday weather as an example. Heat from the sun causes instability as hot air rises, creates convective currents, destabilizes clouds, and moves wind and precipitation around the planet. That wind creates waves that roll across the sea and crash into distant shores, eroding and moving coastlines and ecosystems. This self-assembly creates nature as we know it—the word itself arising from the Latin *natura,* the "course of things." It is what Ralph Waldo Emerson called the "mutable cloud, which is always and never the same."

This vast complexity and mutability was one reason that I had a hard time getting a simple answer out of my sources. There were too many irregularities, too many unknowns. As noted earlier, glaciology wasn't even a discipline until the mid-twentieth century, when M3 was among the first to invent it. Techniques, baselines, measurements, and tools were still being created, many of them right here at JIRP. I was not alone in my race to find answers; everyone at JIRP was on a similar mission.

One thing we do know, Markle said during our chat, was that drivers of climate change are not mutually exclusive. That is, human-caused climate change—which is also very real—takes place alongside natural climate change and even intensifies it. "There are natural oscillations that create eight degrees of warming, and there's human-caused effects that make eight degrees of warming," Markle said. "The climate system itself is a big complicated thing, and it changes for a lot of reasons. The fact that one happens doesn't mean the other can't happen."

An example of the union of natural and artificial climate change can be seen in nine "natural" tipping points in the planet's climate system—such as the flora and fauna die-off in the Amazon rain forest, which will acutely increase carbon in the atmosphere, and the thawing permafrost and its hidden trove of greenhouse gases. After these events, initiated by human causes, the world will likely see unstoppable, catastrophic natural warming. Another gauge of the importance of ice on earth: five of these tipping points involve melting ice. Even if humans fast-track decarbonization to an unthinkably short thirty years and keep to the Paris Agreement target of 1.5 degrees Celsius, we may very likely pass one or all of the remaining thresholds.*

Markle picked up a handful of snow and sifted it through his glove. He seemed lost in thought, as if a new theory about climate change and melting ice were materializing in his mind right then. Five orange and yellow tents framed the northwestern perimeter of the camp a hundred yards away. A wind block of heaped snow surrounded each tent, with cords and stakes lashed down to keep it from blowing over. Markle knelt and used a gloved finger to draw a timeline. "If you look at the development of human civilization," he said, "snow and ice have definitely played a big role. Twenty thousand years ago, it was a lot colder. Ice was everywhere. The exact same people lived then, not much different in terms of evolution of the human mind and everything else. Then ten thousand years ago, it gets warmer, the ice melts, flows into farmland, and they develop agriculture, civilization, all the history we have ever heard of, all the people and wars, love stories, everything. All of our crops were developed in that very narrow temperature range. And the only thing that changed was the climate. The difference was like four or five degrees. I find that sobering: the history of human civilization entirely depends on climate."

Reminiscing about this and wandering in circles while I waited for my flight at Camp 18, I had inadvertently ended up on the rocky outcropping where Balter first told me about her discovery. Signals from past climates were all around—frozen in ice, sealed in rock, percolating

* Of the five known mass extinctions on earth, all were caused by runaway climate change and several were influenced by a thawing cryosphere. In one, 440 million years ago, sudden glaciation followed by an even more sudden melt killed 86 percent of all life on earth.

through the permafrost. I had spent most of my life in the frozen world, camping beneath subzero skies, walking through sun dogs and glowing circumzenithal arcs on 19,000-foot summits at dawn. But I had had no idea about the role the cryosphere played in our own evolution.

A bank of rain clouds pressed in from the west. A glacier groaned. I could not see a single living thing, and yet I no longer considered this glacial valley a negative space. It was alive. It helped form a verdant climatic window in which human civilization blossomed. Ten thousand years. Four degrees. A blink in time in the history of the planet.

A distant clapping sound echoed through the air. It reverberated off Gilkey and Bucher Glaciers. I turned around and saw my flight rise up from Gilkey Trench like a scene from a 1980s action movie. I grabbed my backpack from the dining hall and hustled to the landing pad. The pilot pointed to the front seat this time, and I climbed through the door. There was no time to lose, he said; the storm was here. Two inches of rain and gale-force winds were forecast for the next seventy-two hours. The cockpit shook as it lifted off, and the shanty shacks of Camp 18 fell away as we banked toward the Gilkey and Bucher Glaciers. I levitated out of my seat on the zero-*g* dive, then clung to the armrest as we zipped past the gnarled granite summit of Triangle Peak and what was left of the muddled Herbert Glacier.

We were leaving this world, I realized. Slowly, imperceptibly. We had been leaving it for some time. Not purposefully at first, but now with some determination. We were pushing on the tipping points, daring them to go. It was hard to notice living in the city. While everything changed in the great white north, things at home appeared to stay the same. The elementary school down the block from our apartment. The bodega on the corner. The plane trees that shaded our street. They were unchanged. There was less snow at home, but there were still blizzards. We joked about the end of winter—prayed for the sun to break through in April, melt the snowbanks, and free rivers and lakes of ice. Back there, it was impossible to see the ends of the earth melting. Climate change moved too slowly, like an ocean tide rising a bit less every year.

We'd been leaving this world for two hundred years, in the same way I was leaving the icefield today, calmly sitting shotgun in the helicopter—green forests, blue ocean, and a raft of cruise ships straight ahead. Beneath us, the Taku Towers looked like something from a dream. Half-mile-wide glaciers flowed downhill. Snow clinging to vertical couloirs cut straight down mountainsides. A massive icefall, larger than I could ever have conjured, collapsed as we flew by. I followed the river of ice downhill; it tumbled away from the towers, a plume of snow billowing before it. It would continue down, even after it settled, creeping with the glacier toward the forests, rivers, and Alaskan towns far below, where one day it would finally meet the sea, calve one last time, and morph yet again into something completely new.

The Alps

10

The Great Melt Has Arrived

In the village of Corvara, Italy, the picturesque portrait of winter remained intact. A cold breeze blowing off the 10,000-foot summit of Piz Boè smelled like woodsmoke. A snow-packed, one-lane road wound through a cluster of broad-roofed chalets, past the town well that had been spewing glacial meltwater into a concrete basin for three centuries. The central bell tower was constructed five years after the *Mayflower* set sail for America. The gable ends of the chalets were carved and painted with edelweiss, mountain scenes, and the occasional portrait of an Alpine lass displaying her impressive décolletage. Thick rafters jutting from beneath the wide roofs—which once provided shelter for two or three farming families—harbored droves of skiers who traveled to the Italian Dolomites to seek adventure through the snow-covered peaks, then get cozy around a hearth and sip hot chocolate spiked with home-brewed grappa.*

It was in this valley, about a hundred yards from where I stood, that winter took a great leap from the season of dread to a season of

* The Dolomites were named for Déodat de Dolomieu, who in 1789 discovered the rock and its unique limestone chemical signature—composed largely of dead life-forms that had sunk to the bottom of a primordial sea millions of years ago.

beauty and celebration—laced with mulled wine, five-star bistros built into cowsheds, and rows of shearling-lined sling-back chairs looking up at some of Europe's tallest peaks. Just a couple hundred years ago, people looked at the Alps, and the eternal winter that embraced them, in a very different way. Exposure to below-freezing temperatures could be a death sentence, with little medical means to treat hypothermia, frostbite, or even a common cold. Avalanches wiped out entire villages. Travelers froze to death or vanished into glacial crevasses. The Alps were tentacles of Aristotle's Frozen Zone—the name *Alps* also derived from the Latin word for "white."

Looking up at the mountains will still give you vertigo. They are steep like the Cascades but twice as tall—twisting into sawtooth ridges, arêtes, crests, cols, and summits capped with thick mantles of blue glacial ice. The range is relatively compact, cutting a 750-mile arc from France to Slovenia over an area the size of Kansas. Sixteen hundred peaks reach above sixty-five hundred feet. For most of history, the mountains functioned as a wall for nations, kings, bishops, feudal lords, and dukes who used the harrowing topographic relief as a natural defense. Borders were set in the mountains; castles and fortifications were built high in the foothills. The now-ubiquitous image of fairytale spires and moats against a backdrop of articulated European summits—as seen in Disney movies, on fancy European chocolate bar wrappers, and on about 20 percent of the books and toys that Grey has amassed in her twenty-four-month life—misrepresents the actual interface among humankind, mountains, and winter that in the not-so-distant past was deadly.*

Some valleys in the Alps are so steep they see daylight for only a few hours on a winter day. Summits like the concave Eiger, the wizard's hat

* Let us not forget the actual horrors of the Middle Ages—religious zealotry, plague, war, authoritarianism, torture, and a life expectancy of around thirty-five.

of the Matterhorn, and the blockish massifs of the Jungfrau and Mont Blanc are so vertiginous—and set so close to major populations—they can be seen from city streets a hundred miles away. Unlike most major mountain ranges on the planet, the Alps do not exist on the fringes of society. They hang like ornaments in the middle of one of the cradles of civilization. Necessity, opportunity, and curiosity are powerful things, and, danger be damned, travelers wandered through the mountains at elevations up to ten thousand feet beginning around six thousand years ago—wearing laced shoes and leather pants and carrying figurines carved from wood and knotted string woven from plant fiber. The first signs of life in the range reach back even further: Neanderthal caves date to around 40,000 BC. The first fingerprint of the Western world: paved roads engineered by the Romans in the first century BC.

The Col Alt ski lift in front of us was not ancient, but it was historic. When it was built in 1946, it was the first chairlift in Italy. Cables were slung from wooden towers, and skiers dressed in high-waisted gabardine ski pants and leather boots rode it for a half century. Pensions and four-star hotels cropped up around the lift, along with a network of ski trails cut through the surrounding Swiss stone-pine forest. Restaurants and shops followed, and the beginnings of the $50 billion winter tourism business in the Alps, which would colonize Frozen Zones around the world and change the way people looked at winter, took another step forward.

Our group standing by the Col Alt that morning was well versed in winter travel and life. My old friend and colleague David Reddick had been the photo editor of *Powder* magazine for nearly three decades. Professional skier Christina Lustenberger, or "Lusti" as friends called her, was a protein-consuming, early-to-bed "silent assassin"—Reddick's words—who had made a name for herself as an accomplished ski mountaineer. She and the fourth member of our party, pro skier Giulia Monego, had been planning a Dolomites

The first winter tourists in the Alps traveled in stagecoaches fitted with sleigh runners. Grown men took up the sport of sledding as a semiprofessional pursuit, organized into clubs with regional and international races presided over by princes and dukes. Ice skating evolved into a kind of outdoor happening, with pavilions, bleacher seating, and a soundtrack performed by a full orchestra. The final chapter in this progression was conveyed on contraptions like the Col Alt chairlift, which ferried thousands of winter tourists a few hundred feet up the mountain so they could try their hand at the fashionable new sport of "skee-ing."

tour that luckily coincided with ours. Monego grew up in Venice and raced on the Europe Cup circuit in and around the Dolomites when she was younger. She was now a professional guide and would, thankfully, double as a translator for us.

Our mission was to document Alpine winter life for a week in February, deduce how it had evolved, and project what would happen to it when winter was gone. Our itinerary was to ski the Sellaronda—a twenty-five-mile circle of lifts, trails, and villages in the Dolomites—over five days. We would ski and hike from village to village during the day and sleep in on-mountain *rifugios*—staffed mountain huts—at night. An indication of the level of hospitality the Alps had attained: porters would deliver our luggage every night to the foot of our beds.

The Alps were a good case study. The mountain range is one of the fastest-warming ranges on the planet—thanks to the previously discussed thermohaline current and interconnected atmospheric currents that are superheating the Alps three times faster than the global average. The effects have been profound. In addition to losing half its glacial ice since the dawn of the industrial age, the range has lost another 20 percent between then and 2000. And the rate of melt is increasing. The European Geosciences Union released a report in 2019 predicting that 90 percent of the Alps' present-day glacier volume could be gone in eighty years. Lower-elevation mountains, like those in Austria, have it worse, warming 3.6 degrees Fahrenheit since 1880. Snowfall in the iconic French ski town of Chamonix has been cut in half between 1980 and 2020, and the ski season across the entire Alps has shrunk by a month since 1960, forcing more than two hundred ski resorts in the range to close already. The *Cryosphere,* a journal of the European Geosciences Union, predicted 70 percent less snow in the mountains by the end of the century.

The Southern Alps were warming the fastest, with precipitation decreasing, the snow line rising, and invasive species like palm trees now appearing in various forests. The last living glacier in the Dolomites—the great Marmolada Glacier—sat in the middle of our ski tour. The wall of ice, draped on an eponymous 10,968-foot peak, has been visible from the canals of Venice sixty miles south for thousands of years—through the doges, the Crusades, the plague, the Renaissance. It watered farms and vineyards in the 1,500-square-mile Piave River catchment for equally as long. But according to researchers, the glacier will soon be gone. A 2019 study published by Italy's National Research Council found that the Marmolada shrank by a third and saw its surface area reduced by a fifth in just eleven years, between 2004 and 2015. This period includes the relatively chilly winter of 2013, after which headlines proclaimed that the queen of the Dolomites had stopped melting. (It resumed its retreat that summer.)

Some of the dominoes that Riedel, Gleason, Campbell, and others had alluded to throughout my journey—rapid melting, alpine habitat degeneration, drought, decreased albedo, and crop failure—had already started falling here. In the remote forests of the Cascades and the barren icefields around Juneau, the end of winter seemed like an abstract theory. Here, in the Alps, where thirteen million people woke each morning and looked outside at a pure white snowscape all winter long, the Great Melt was vastly more visible.

Our guide, Marcello Cominetti, did not seem concerned with the nuances of our quest. He wanted to get lunch. It was 10 a.m. We had yet to put on our skis. Cominetti was one of the best climbers, skiers, and guides in a valley known to breed some of the world's great alpinists.* He'd been guiding for more than forty years. He scaled a climbing route on Monte Fitz Roy in Patagonia with a client for fun, back when only a handful of people had completed it. When Sylvester Stallone and a film crew of two hundred arrived to film the ludicrous, preposterous, and wildly entertaining climbing film *Cliffhanger,* they hired Cominetti to be a stunt double. Stallone needed it, Cominetti said. "He step out of the helicopter the first day in his loafers!" he exclaimed. "He slide two hundred meters and almost die!"

After a quick gondola ride, we found a midmountain bistro set on the edge of a snowfield. The building was hand-hewn—thick

* The first climbers to summit K2 in Nepal came from this valley.

pine rafters, pitch-stained siding. Inside was a small kitchen, a rack of wine and liquor, and a cash register. Most customers were trying to get out as we elbowed our way in. A windstorm was coming, Cominetti said. Maximum gusts could hit 125 miles per hour—well beyond the threshold of a hurricane and into the realm of tornadoes. Outside, a hundred skiers frantically duck-walked to the lift to get a ride down. Cominetti was more worried about the quality of the barley soup that he had ordered. "This is not food," he said, shrugging his shoulders. "I should call the polizia."

Marcello Cominetti and Giulia Monego skinning up the first pass. More frequent violent windstorms are a new phenomenon in the Italian Alps. Another recent storm had flattened a nearby forest where Antonio Stradivari, the seventeenth-century luthier, had harvested wood for his instruments. Several recent studies concluded that another climate event—sudden cooling in the 1600s—is the cause of Stradivarius instruments' unique sound. Nearly a century of frigid winters caused slower growth in the trees, making the wood vastly denser and thus acoustically distinct. (The violins were put through a CT scanner at Mount Sinai Hospital in New York City during the study.) (David Reddick)

There was no need. The polizia were already outside driving snowmobiles in circles with blue flashing lights, trying to corral the mass hysteria unfolding on the snowfield. Gusts of wind sounded like eighteen-wheelers rushing past on a highway. The lift we had just disembarked from shut down; the one we were supposed to ride next closed five minutes later. Without a word, Cominetti paid his bill, walked outside, put climbing skins on his skis, and started walking uphill. It was one in the afternoon; it would be dark in four hours. Monego, who had skied the Dolomites since she was a child, filled us in. We had to cross two mountain passes to get to the first rifugio. All the lifts were now closed. Cominetti disappeared into a cloud of swirling snow as we applied climbing skins to our own skis. Then we lowered our heads and took off against the exodus, straight into the teeth of the storm.

Demons of Antiquity

Like many unexplained phenomena in antiquity, winter and mountains were often associated with deities and demons. (See the gods of Mount Olympus, the "pests" of Mount Kippumaki in Finland, Buddha's footprint on Adam's Peak in Sri Lanka, and the mythical Mount Meru.) In Europe, Mount Canigó in the French Pyrenees sheltered evil spirits in a castle set at the bottom of a lake—the spirits would stir up violent storms if disturbed. Medieval miners lived in fear of Montagnards, demons that dwelled inside the mountain. The giants of Kohlhütte and Albach mountains in the Tyrol region could smell human flesh from miles away, and in Austria, the Alpine giant Heimo haunted the people of Innsbruck—until he was baptized and built the bridge over the Inn River as penance.

Rock formations known as Devil's pulpits,* where Lucifer him-

* Look up Jakob Götzenberger's surreal nineteenth-century frescoes of Devil's pulpits at Baden-Baden in southwestern Germany.

self preached, were mapped and inspected by bishops and their minions.* Images of ghouls lurking in the mountains were embossed on leather book covers of the Hapsburg empire, and evidence of witches and dragons was meticulously collected and distributed across the continent throughout the Middle Ages.† Villagers in Switzerland's Lötschental Valley still wear demonic-looking costumes, called Tschäggättä, every February to pray for the end of winter. Tales and folklore from the philologists-anthropologists-lexicographers Brothers Grimm in the early nineteenth century popularized many Alpine tales about the dangers and demons of winter. As Jacob Grimm wrote in 1811, "In the high mountains, in self-contained valleys, there still lives a timeless meaning at its purest; in narrow villages, with a few paths and no roads, where no false enlightenment has arrived or carried out its worth, there still resides a hidden store of national customs, legends, and faith."‡

I found some examples of prehistoric Alpine civilization at the Historical Museum of Bern during a layover a few days before the ski tour began. The museum was not hard to find—nothing is in Switzerland, with its shiny red-and-white directional signs, painted pedestrian and bicycle paths, and free maps. A very old soul sat next to me on the city bus that morning. He said that Bern—itself a

* The missionaries' last stop was the town of Gernsbach in Germany's Black Forest, where foresters refusing to take the sacrament beat them back.

† Maria Savi-Lopez's 1889 *Leggende Delle Alpi* (Legends of the Alps) (dedicated to the Queen of Italy) recounts the dragon lore that pervaded the Alps in the Middle Ages. Eyewitnesses, unidentifiable bones, and a study of dragons' flight and habitat suggested the existence of the mythic beasts and kept people away from the mountains.

‡ This glowing description is a departure from the nightmarish tales the brothers gleaned from the area and passed on to the children of Europe. For example, in the Grimms' original *Snow White,* the queen orders her huntsman to kill Snow White—her biological daughter in the account—and bring home the child's lungs and liver so that she can eat them.

gateway city to the Alps with views of the Jungfrau and Eiger—was founded in 1191. Then he pointed out a few of the city's early structures as we drove by: the fifteenth-century town hall and the Zytglogge, a twelfth-century clock tower with spooky mechanical puppets that call out the hours.

The bus rumbled over the Nydeggbrücke bridge that spans the Aare River, and I stood up to get off at the museum. "You know this museum is not old?!" he exclaimed. He was right. The museum was built in 1894, practically yesterday for the Bernese and quite possibly the year of my new friend's birth. It did hold some of the oldest artifacts in the Alps. Twenty feet inside, I stumbled across an exhibit of the first hominids to arrive at the foot of the mountains—some 250,000 years ago. Ice ages were still coming and going then, wiping the landscape clean with vertical miles of ice, then exposing it again. Around 22,000 BC, ice filled almost every valley in the range; twelve thousand years later, it receded and hunter-gatherers moved in. As the ice retreated, it uncovered a barren landscape with little vegetation. Willows and dwarf birch took root first; wild game, horses, reindeer, and boars followed. The first settlers lived in caves and used animal skins and feathers to make clothes. Bones and antlers were shaped into sewing needles, knives, and other tools. Amulets and jewelry made from black jet, lignite, and amber, likely traded from the Baltic region, appeared around the same time the tribes migrated into the mountains.

As Minoans, Mycenaeans, and Phoenicians roamed southern Europe during the Bronze Age, dozens of groups such as the Celts, Lepontii, Gauls, and a confederation of tribes known as Rhaetians settled down in the Alps. Many Rhaetians, probably migrants from Etruria, vacated Rome shortly after the last Etruscan king of Rome was deposed.* Roman soldiers eventually arrived at the foot of the

* Note this king's awesome name: *Lucius Tarquinius Superbus!*

mountains during the Alpine Wars. The mountains were lawless then, a place of anarchy, barbarism, and micronations with such lords as Luerius, the king of the Arvernians in the Auvergne region of France. Tales recount how he traveled with a parade of clansmen, huntsmen, and packs of hounds—Luerius safely seated in a silver chariot, tossing gold coins to villagers along the way. The Romans fought dozens of similar groups, such as the Vocontii Gauls of Vaucluse, the Allobroges in the Isère Valley, the Ligurians, and the Salyes, until they finally declared conquest of the mountains in 14 BC.

Rhaetians living in the Dolomites did not fight the Romans but rather joined ranks and fought alongside them. When the war was over, the Rhaetians settled down with their conquerors. In a few short decades, a mash-up of the vulgar Latin the soldiers spoke and the Rhaetian tongue blended into Ladin—the language of the five valleys we would be skiing through. Ladin is still taught in elementary schools, and 80 percent of families in the five Ladin valleys use it as their first language at home. The circuit we planned to follow was an ancient trading route Ladins used to pass through the valleys: Val Gardena, Val Badia, Val di Fassa, Livinallongo, and Cortina d'Ampezzo. The region existed in the Austro-Hungarian Kingdom until Italy annexed it after World War I. Now it is a self-governing province, a *liberi comuni,* within Italy, and one of the oldest mountain cultures in the world.

"It Is My Only Love"

Clouds streamed over steep ridgelines dividing the Ladin valleys as we neared the top of the first pass. I still had no idea how my knee would hold up on an actual ski run. Skinning had been painless so far, though every step felt wobbly. The fluffy corduroy that the resort's grooming machines left on the upper slope was now wind-hardened ridges, reducing the surface area of our climbing skins by 50 percent. With half of the traction gone, we all started slipping—a dangerous and awkward situation on a steep slope, in free-heel bindings with no control over the ski and a thousand-foot drop-off on the right-hand edge of the trail.

The wind ratcheted up to biblical strength the closer we came to the top. No one spoke for an hour. Every step was precarious, until a final gust, thankfully from the rear, blew us up the last pitch to the summit station. Milky clouds pushed down on the peaks—Cristallo (10,549 feet), Tre Cime (9,850 feet), Gruppo dei Cadini (9,312 feet). Swirls of snow blinded us, followed by splotches of blue sky that momentarily opened up. There were no skiers, no ski patrol, no lift attendants. The deserted scene was beautiful and haunting, and I wondered if this was what ski areas would look like in fifty years—when there wasn't enough snow for a resort to stay open but still enough to ski on.

Cominetti stood with his back to the wind, looking out over the expanse of peaks. The fugue state he seemed caught in was endearing. He was a model Alpino—a term that refers to the mountain infantry of the Italian Army. The Alpini had single-handedly turned the tides of both world wars, with unreal mountaineering feats like climbing vertical ice for days with rifles, mortars, and explosives strapped to their backs. Cominetti joined the division during the compulsory year every Italian male spends in the military. He was a natural at nineteen years old and made officer quickly—which meant he got fiberglass skis and plastic boots instead of wood and leather. New recruits were not so lucky. They had to prove their skills on the steep slopes around Corvara on eight-foot wooden skis with no edges, often just once when commanding officers arrived for muster. All of which made for a chaotic mess, Cominetti said, as novice troops bombed through the woods, eyes closed, hands in front to absorb a tree or rock to the face, all of them muttering rosaries, trying to make it to the bottom where a line of brass awaited with clipboards in hand. During the halcyon, neon-colored detente of the 1970s, when Cominetti was coming up, one of those soldiers was Italian ski racing sensation Alberto Tomba. That year, to lighten the mood and inspire its mountain troops, the military held a ski race for the soldiers, and the winner would get a new car. Tomba did not win, nor did Cominetti, a testament to the talent in the Alpini division at the time.

In the week I spent with him, I saw that Cominetti was also a dreamer, a Renaissance man who made his money on skis but might have been happier playing music or discussing philosophy. He often asked me about film, literature, and politics and shared his many theories on human existence. "There are two kinds of people," he told me that morning, apropos of nothing. "People who fall in love with a mountain guide and people who fall in love with a ski instructor." I might have hit the proverbial nail on the head when I asked him if all of his passions could be wrapped up in the field of anthropology. "It is my only love," he said.*

Cominetti took off down the slope without a word. We followed through a maze of ski trails, empty chairlifts, bridges, lift stations, restaurants, and roads. At the bottom of the hill, he stopped next to another stationary chairlift, put his skins back on and started up the second pass. The wind continued to build, making the corduroy icier and the skinning even more difficult. It was almost dark; the rifugio was hours away. We climbed like the Alpini, in a line, everyone taking extreme caution with each step and nearly falling every

---

* To get a sense of the artist who occupies his soul, look up his band, the Frozen Rats, which is composed of Cominetti (a guide) and his friend (a ski instructor). They specialize in extended Pink Floyd covers and pop-up performances, often beneath the thin plastic shelter of a pavilion tent. The music is mostly slow with melodic guitar solos over a synthesizer-drum machine. Interpretations of "Dark Side of the Moon" can last up to twenty minutes.

twenty feet. Leaning into the wind, I found it easy to see how winter could instill terror in ancient populations. We wore layers of wicking microfiber, plush down jackets, and waterproof, breathable Gore-Tex shells. The first climbers in the Alps—who wore animal furs, linen, and wool—would have likely frozen to death in a storm like this.

One such victim was a man known today as Ötzi the Iceman. Ötzi was disentombed from a glacier forty miles northwest of the Sellaronda around five thousand years after he lay down to die. Two German hikers found the corpse in the Ötztal Alps as they were descending Finail Peak in the Tisenjoch region. They spotted the half-buried body leaning against a flat rock at 10,500 feet. The cadaver was almost perfectly preserved, along with much of its clothing and possessions. Forensic experts and archaeologists dated Ötzi to between 3350 and 3100 BC and deduced that he was a man of some means. Studies of nearby stalagmites, fossilized tree rings, and ice indicate that the Alps were mostly ice-free at the time. Ötzi was around forty-six years old when he died. He wore a woven grass cloak, a leather loincloth, and a bearskin cap with a chinstrap. He was armed with a dagger, a bow, a quiver of dogwood arrows, and a highly valuable copper-tipped axe. A pouch sewn into his belt kept flint flakes and fire-starting material dry. He had been shot in the back with an arrow and, after gathering his personal possessions around him, including two rolled pieces of birch bark, he sat down in the snow and died. He had not gone easily. The blood of four other men was found on his clothes.

It was too windy to talk at the top of the next pass, so Monego pointed down at a little shelter clinging to the side of the mountain. We switched our touring bindings to skiing mode for the last time that day and pushed off again. There wasn't a single ski track on the trail. We carved long S's under the empty chairlift swinging overhead, already envisioning warm beds and a hot meal waiting inside.

At every turn, I felt my quadriceps and knee fall into sync with the slope. Angulate, lean in, let the ski edge bite through the turn. The rush of speed caught me off guard after a year off skis, as if a gust of wind were heaving me down the slope.

I thought about the Nick Adams stories that Ernest Hemingway wrote about skiing not far from here. His favorite resort was Austria's Silvretta Montafon, about a hundred miles northwest of the Sellaronda. He, his first wife Hadley, and their son Jack checked into the Hotel Taube at Schruns around Thanksgiving in the 1920s and didn't leave until Easter. Hemingway spent most of his time drinking kirsch and gambling with, as he later wrote, "country men in the Weinstube with nailed boots and mountain clothes."

The Hemingways liked to skin up the Bielerhöhe at 6,600 feet and stay in the Madlenerhaus overnight, then ski down the following day.* In a picture of the couple in Schruns, Hemingway is so skinny it looks as if his wool pants are about to fall off. He is wearing a ski sweater with a single stripe around the waist, a matching hat with a small pom-pom, and a patchy beard. Hadley looks ebullient, healthy, and tanned. Jack sits in her arms, squinting into the bright sunlight, an Alpine scene layered with thick snow behind them. Hemingway was an expert skier by then and his descriptions of skiing in the Nick Adams's stories are so well rendered that many aspiring skiers at the time used them as instruction:

> On the white below George dipped and rose and dipped out of sight. The rush and the sudden swoop as he dropped down

* An Austrian ski instruction manual titled *Lilienfelder Skilauftechnik* by Matthias Zdarsky—a sculptor, painter, philosopher, and militant health nut who developed the first steel ski binding and proper "stem turn"—popularized skiing in Europe. (Zdarsky's skis were nine and a half feet long and weighed ten and a half pounds each.) By the 1920s, thousands of skiers had schussed through the Alps, skiing had become an Olympic sport, and the modern era of winter living had begun.

> a steep undulation in the mountain side plucked Nick's mind out and left him only the wonderful flying, dropping sensation in his body. He rose to a slight up-run and then the snow seemed to drop out from under him as he went down, faster and faster in a rush down the last, long steep slope. Crouching so he was almost sitting back on his skis, trying to keep the center of gravity low, the snow driving like a sand-storm, he knew the pace was too much. But he held it.*

We skied to the front door of Rifugio Baita Cuz, clicked out of our skis, and walked into a warm saloon. Reggae played on the stereo. Drying racks covered one wall. There was a boot-drying station downstairs. A bartender asked for our order, and by the time I took off my jacket and goggles, Monego had handed me an Aperol spritz. Perhaps it is these extremes—precipitous heights and violent storms followed by biodynamic wine and tagliatelle with venison ragù—that makes modern winter living in the Alps so extraordinary. Our double rooms were outfitted with fluffy comforters folded at the foot of each bed. Through a small, paned window the wind howled around steep, rimed peaks. The duffels we had packed with slippers, sweatpants, toiletries, and books had been delivered by snowmobile earlier that day.

---

* Quite a few writers preceded Hemingway into the Alps in winter, effectively popularizing the destination to their fans. Jean-Jacques Rousseau was one of the first, with his writings about the tranquility and purity of the Alpine lifestyle. Mary Shelley attracted thousands to the Chamonix Valley with terrifying passages about the Mer de Glace glacier in *Frankenstein*. Robert Louis Stevenson spent the winters of 1880 and 1881 in the Swiss Alps, writing mostly about sledding and other winter sports. Friedrich Nietzsche, Hermann Hesse, and Thomas Mann all wintered in the Alps, Mann in Davos in 1912 and 1921, inspiring his novel *Magic Mountain*: "But if there was something roguish and fantastic about the immediate vicinity through which you laboriously made your way, the towering statues of snow-clad Alps, gazing down from the distance, awakened in you feelings of the sublime and holy."

I took a hot shower, then scrambled across the frozen deck to a Finnish barrel sauna looking out over Val di Fassa, the southernmost Ladin valley. A naked German couple awkwardly rearranged themselves to fit me in. Two frosty port lights looked out at a silhouette of mountains through a mist of airborne frost. Wind gusts pushed through the fitted cedar planks and rocked the barrel just enough to imagine rolling a mile downhill with an assemblage of German body parts and red-hot sauna rocks.*

That night at dinner, I indeed ordered the tagliatelle with venison ragù. Cominetti held the room with a treatise on Luis Trenker, a local filmmaker from the 1930s who hit it big with a 1932 Universal Pictures movie, *The Doomed Battalion,* set right here in the Dolomites. This was no *Cliffhanger.* The movie followed actual exploits of the Alpini and Austro-Hungarian mountain infantry in the Dolomites during World War I—with Trenker acting, skiing, climbing, and performing many of the film's stunts. The highlight: a graceful, high-speed powder run down the Marmolada, in which waves of silvery snow arc across the black-and-white screen while Tyrolean soldiers fire machine guns at him. *Spettacolare!* Of particular note for Cominetti, Trenker had been an architect first, then a bobsledder

* A fascinating side history of winter tourism in the Alps is the so-called medical tourism that preceded and ultimately created it. In the opening decades of the Industrial Revolution, European city dwellers sought refuge and healing from a proliferation of pollution—and resulting tuberculosis. Doctors in the Alps built *Kurhäuser* (health spas) for "clean air" and "milk" therapy, the latter requiring patients to drink a gallon or more of milk every day to absorb vitamins and minerals from Alpine pastureland where livestock grazed. "Dung therapy" came next, in which doctors administered juice from pressed cow dung—sometimes mixed with wine and spices—to cure respiratory ailments. Ashes from burned dung were given to guests with dropsy, and poor souls with an earache had ox urine mixed with myrrh delivered to their bedside. After a few *Kurhäuser* in Davos and St. Moritz stayed open through the winter, patients were encouraged to sled, skate, and, eventually, ski to help circulation—giving rise to the first skating rinks, ski lifts, and winter tourists in the Alps.

and an alpine guide, then a writer, a director, and an actor, in that order. "This man!" Cominetti exclaimed. "He is my hero!"

That night I stepped onto the bedroom balcony and looked up at the shadow of the Rosengarten Massif. Clouds passed at a hundred miles an hour, dragging dark shadows across the valley. Pale blue starlight between clouds lit the hills—until another cloud crested the ridge and carried the mountain back into night. It seemed impossible that we could be so comfortable in such an inhospitable place.

Monego descending a slope in front of the Marmolada. The first modern exploration of the high peaks came as gentlemen mountaineers of the Enlightenment sought to explain natural wonders across the continent, like glaciers and waterfalls, typically with a few cases of wine in tow and a brace of pistols to "test the acoustics" of chasms and geological marvels. The intrepid climbers of the golden age of Alpinism came next, conquering the most difficult routes in Europe using Manila hemp rope, leather boots, forged-steel crampons, wood-splitting axes, and a metal-tipped alpenstock staff. What followed was an ever-widening flood of Alpine tourists—a flood that grew from a few hundred to a few hundred thousand in the second half of the nineteenth century. (David Reddick)

The scene reminded me of late-night walks through the ancient streets of Engelberg, Switzerland, and villages in Verbier, where friends overwintered in tiny farm shacks. I thought about partying in Chamonix after skiing the terrifying steeps of the Aiguille du Midi, then stumbling home past the little graveyard on Avenue Cachat Le Géant—where the valley's fallen guides were buried—and its chapel, with skiers' names etched into the stained-glass windows. We had slept in tents in the Himalayas, Andes, and Rockies. In the Alps, the summits had their own hotels, affixed with steel anchors drilled into the rock.

Years ago, I spent a week at one of the first Alpine refuges ever built. You can't drive to the thousand-year-old St. Bernard monastery in the Swiss Alps. You have to park and skin up St. Bernard Pass.* My host and guide then was a septuagenarian treasurer from the Vatican. Jean-Marie escaped from Rome to the hostel whenever his bosses would allow. He was a monk, the irreverent Friar Tuck type, who liked to crack jokes and make fun of his brothers. He hiked around the monastery on summer vacations, when edelweiss and génépie noir cover the slopes, and he skied there every winter, talking with nature and God and carving wild, erratic turns down Pointe de Drône, Grande Chenalette, and Mont Fourchon.

Jean-Marie learned to ski when he was studying at seminary nearby in Martigny, Switzerland. In the winter, he and fellow monks walked into the hills once a week with plastic garbage bags in their pockets. When it was steep, they sat on the bags and slid on their

* The Great St. Bernard hospice was built in 1050 after an archdeacon from Aosta, Italy, named Bernard de Menthon dedicated the structure to the protection of St. Nicholas of Myer and assigned a community of friars to operate it. He gave them a motto: *Hic christus adoratur et pascitur,* "Here, Christ is adored and fed." The monastery was the first mountain rescue station in the world, from which the world's first avalanche rescue dog, the lumbering St. Bernard, would soon arise—bred, trained, and managed by monks.

butts; when it wasn't, they flopped onto their bellies and went head-first. "There was quite a technique to it!" he exclaimed. I skied with Jean-Marie every day for a week. Skinning up, he would chant Hail Marys and tell me jokes about Belgians. On the way down—typically around halfway—he would stop turning altogether and run straight for the bottom, arms spread wide, wisps of snow flying from the tails of his skis, hurtling completely out of control at thirty miles an hour.

When we weren't skiing, I explored the tubular crypt in the basement where church services were held and the thousand-year-old cheese cellar hidden next door. Simple meals of pasta, grains, and bread were served at long banquet tables, where a few dozen mountain guides, churchgoers, climbers, and skiers sipped bowls of steaming coffee in the morning and tiny juice glasses of blended red Valais wine at night.* (The hostel can host up to 160 guests per night.) I fell asleep staring at snowflakes meticulously carved into the rafters of my room. By the time I got to the mess hall every morning, Jean-Marie was waiting in his baggy white windbreaker and 1960s ski boots. One afternoon, as we took off our boots in the ski room—which was paved with Roman millstones—I asked Jean-Marie his secret to appreciating the solitude and beauty of winter life. "Do what the Japanese do," he said. "Take lots of pictures."

* One famous guest was Napoléon Bonaparte, who in May 1800 crossed the Grand St. Bernard with forty thousand troops. Among other items on his transit bill—which was never fully paid—were 21,724 bottles of wine. (The monks still send the French government a bill every year, with interest.)

11

The Big Picture

Part of the writer's journey that is rarely acknowledged is the black hole an author enters about halfway through a book when tying the threads of a story together. The meat of the narrative—characters, backstory, scene, hook, plot—is such a mishmash of anecdotes, discoveries, locations, loose ends, and analysis a writer can rarely foresee how everything is going to line up until months or years into the project. I was deep in this black hole, about halfway through reporting for this book, when I received an email that began pulling things together. I was sitting in the studio that Sara and I had built into a hundred-year-old warehouse we were renting, staring at cobwebs growing from the ceiling. It was winter. Muddy, brown snow covered the sidewalk. A cold draft blew through Plexiglas panels that the landlord had screwed to three window frames shortly before we signed the lease. I had stopped maintaining the place months before—too deep in the process of untangling this story to think about anything else.

I had already visited Juneau and the Cascades. The fate of the world's great ice sheets and glaciers was clear. I was still obsessing over what would happen to us—in Brooklyn, in my childhood home in Maine, at our cabin in the Catskills. What about water itself and developing nations with few resources to mitigate water shortages, flooding, drought, deadly heat waves, and disruptions to food production? I emailed every snow scientist I had ever met, including a good portion of the JIRP faculty. With every nonanswer came a suggestion to email a man who might be able to help: Michael Zemp.

Dr. Zemp responded that day. His email signature was inspiring: Director of the World Glacier Monitoring Service. I set up a meeting to swing by his office that winter, and two months later, I pulled into an automated parking garage at the University of Zurich. It was midmorning, and I immediately got lost in the modern glass-and-concrete structure. With its vaulted ceilings divided by a 200-foot-long skylight, the main hall of the campus looks more like an upscale food court than a college. After inadvertently backing into a fast-moving line near a cash register, I realized that it *was* an upscale food court, and two dozen hungry scholars were about to toss me out if I didn't buy something or get out of the way.

Zemp found me a few minutes later huddled next to a ficus tree. A Bauhaus-inspired clock overhead read ten thirty in the morning. He was wearing jeans and a navy blue crewneck sweater. His tussled brown hair and the scruff around his goatee suggested a man with little free time. His charges include two young boys, a small office staff, and one of the largest cryosphere observation programs in the world. He showed me the fruits of his labor a few minutes later on a large flat-screen monitor hung in the hallway. This was the battleship control center I'd expected to see at JIRP. The high-definition world map was mostly green and blue. Splotches of magenta and white indicated the planet's ice inventory. Zemp swiped and pinched the screen, as you would on a smartphone, spinning the earth with a flick of his index finger and zooming in and out of glacial valleys and icefields. He spoke rapidly, with urgency, as if there were something happening here but people weren't tuned in to it.

Here were the Cascades, where Jon Riedel had been sending glacier monitoring data since the 1990s. Here were Kelly Gleason's scorched forests. Here was Seth Campbell studying Alaska. Zemp knew them all; he was indeed the central processing unit of the glaciology world. Every scientist I had met so far was laser-focused on the ice they were analyzing, but Zemp saw the big picture, which was looking more and more grim.

As a student, Zemp had wanted to study medicine and geography. He applied to universities specializing in both, but after reading about the launch of a major new earth-observing satellite called Envisat—designed to collect data on things like snow, ice, atmospheric chemistry, ocean temperature, waves, and hundreds of other data points—he decided to merge his interests and study the health of the planet. His first subject as a glaciology student was the Findel Glacier near Zermatt, Switzerland, one of the largest glaciers in the Alps. Measuring it in situ, or on-site, with the help of Lidar lasers installed in an airplane—capable of shooting twenty thousand pulses a second—the team he was with discovered that the glacier had lost forty-nine million cubic meters of ice between 2005 and 2010 and a quarter of its surface area since 1850. It was a shocking revelation: he saw it all, how the altered atmosphere was melting Europe's iconic glaciers more than anyone thought possible. It was on the scale of climate events he'd studied in school—planet-altering events that caused mass extinctions and created new epochs.

The great tragedy about winter and snow in the world's mountain ranges is threefold: the climate is warming faster (1) at higher elevations, (2) at higher latitudes, and (3) in the winter. In the Alps, nearly every glacier under 11,500 feet is predicted to vanish in the next twenty to thirty years. If governments stick to Paris Agreement targets, that melt could be contained to just two-thirds of the ice in the Alps, but the odds of meeting those goals look ever slimmer.

For many remote parts of the world that Zemp monitors, the end of glaciers, snow, and winter will initially pass without protest. Winter rainfall, brown summits instead of white ones, and dying boreal forests—hundreds of miles from major cities and populations—won't incite protests from the masses. But in the Alps, where every peak, valley, draw, and ridge harbors a village, rifugio, resort, highway, lift, hiking trail, or grazing pasture, every tenth of a degree is noticeable. "We are one of the densest-populated mountain regions," Zemp told me. "We really live in the mountains. We have a lot of infrastructure, all our transport means, the trains, the highways, the electricity, the water pipelines, it all goes through the mountains."

In the last ten years, shorter winters and extreme droughts in Europe—with less meltwater to buffer them—have dried up several of the continent's major watersheds, he said. Europe is lucky that in many regions, an increase in precipitation—remember that a warmer atmosphere holds more moisture—will likely fill in for much of the water that melting glaciers no longer provide. Except when it can't. The massive Rhône watershed gets most of its water from melting snow and ice and supplies 150 million people in four major basins. With river flow predicted to decrease in the hot, dry summer months, this vital water source could shrink significantly.

Hydroelectric power—which accounts for 10 percent of the European Union's power generation and 17 percent of the world's electricity supply—often relies on meltwater from snowy mountains and has already lost substantial production. Around 70 percent of Austria's electricity comes from hydropower. There are nineteen hydropower plants on the Rhône River alone. Some 25 percent of Switzerland's hydropower comes directly from glaciers. (In Iceland, that number is 91 percent, and in Norway, 20 percent.)

With the glaciers, or frozen water towers, of the continent drying up, Europe is facing other issues that no one expected, Zemp said. Sediment transport, biogeochemical and contaminant fluxes from rivers to oceans, and biodiversity in riverine and near-shore marine environments—which could affect the food chain and the fishing industry—are all changing in ways scientists do not completely understand. A drought in 2019 dried up and blocked shipping lanes on the Rhine River—where glacial meltwater has dropped 28 percent between 1973 and 2010—significantly slowing economic growth across all of Germany in the third and fourth quarters. "Look at the extreme year of 2003, where we had this massive heat wave," Zemp said. "The glaciers lost up to two meters, more than twice the current melt rate, and then it was also a very dry summer, about three months without rain, a lot of radiation, very strong heat. The problem was that there wasn't enough water in the rivers, which was mainly coming from glaciers, for shipping freighters. The airport then ran out of fuel that was typically delivered by freighters. Then nuclear power plants had to quickly reduce production because the river water they use for cooling reactors wasn't there. There are a lot of consequences that we did not think about before."

With data on climate change evolving every day and internal feedback loops altering predicted outcomes constantly, the entire world appeared to be orbiting a black hole of potential storylines. Even in Europe, home of a quarter of the world's wealth, researchers couldn't build their way around the ramifications of winter's end. Another bizarre side effect that people in the Alps were dealing with was this: many of the high peaks are held together by permafrost—ice crystals that bind loose rock. As the permafrost thaws, entire mountainsides are falling apart. Ground temperatures in many parts of the range are warming five times faster than are air temperatures, threatening nearly every tunnel, highway, ski resort, and adorable Alpine village. This is not a forecast; it is already happening. In 2018, two of the most popular rock-climbing routes in the Mont Blanc area—Bird's backyard—fell to the ground. In July 2006, twenty million cubic feet of rock—half the size of the Empire State Building—broke off the Eiger's east face.

In small Swiss towns like Guttannen in the Bernese Alps, villagers now live in constant fear as millions of pounds of rock directly above them are poised to collapse. In August of 2020, homes near Courmayeur in Italy's Aosta valley were evacuated after more than seventeen million cubic feet of ice threatened to slide off the Planpincieux Glacier in Grandes Jorasses park. (A group called the Safe Mountain Foundation monitors two hundred glaciers in the Aosta valley region alone.)* Even international borders, set on glaciers, are moving. The Theodul Glacier at Testa Grigia above the Swiss mountain resorts of Zermatt (in Switzerland) and Cervinia (in Italy) is thinning, bringing territory and the popular Rifugio Guide del Cervino into Switzerland. A commission has been appointed to study the issue and come up with a possible land swap to keep the rifugio in Italian hands.

Several ski resorts, such as Pontedilegno-Tonale, not far from the Marmolada, have started to cover their glaciers with white geotextile tarps to slow the melting. Just to the north, Davos and the French resort of Courchevel stockpile snow midwinter under heaps of wood chips so they can use it later in the spring. The goal is not always to save a ski trail. Glaciers insulate permafrost, essentially keeping the top of a ski resort and all the lift stations, restaurants, glass-sided hotels, and panoramic viewing stations in place. When the permafrost thaws, everything could come tumbling down. In one case in 2004, two Swiss resorts hired consultants to analyze the melting permafrost on their peaks. When the firm found that, indeed, the mountains were thawing and might fall apart—taking lifts, buildings, and possibly skiers with them—the resorts ordered the study stopped and the results suppressed.

Zemp often skis with his boys and, like me, worries that they will

* A similar hanging glacier collapsed in India's northern state of Uttarakhand in February 2021, killing dozens of people and blasting through two hydroelectric dams.

not be able to enjoy winter in fifty years. More than sixty million tourists visit the Alps every year, generating $30 billion in the ski industry alone—which employs more than 10 percent of the region. The rate of warming in the Alps has increased to 0.9 degrees Fahrenheit per decade since the 1980s, pushing the upper end of temperature predictions for the region to 9 degrees Fahrenheit by 2100. Besides a shorter winter season and winter rain instead of snow, the snow line itself is moving uphill five hundred feet for every degree Celsius of warming—putting at least half the ski resorts in Switzerland out of business, and far more in Germany, Austria, and the Pyrenees.*

The situation at my home mountains in the United States was not much better. Winter season lengths are projected to decline at ski areas across the nation, in some locations by more than 50 percent in the next thirty years and by 80 percent by 2090 if greenhouse gas emissions continue to rise at their current rate. Average temperatures in south-central Colorado have risen by 2 degrees Fahrenheit since 1988. In California's Lake Tahoe region, home to more than a dozen ski areas, warmer temperatures since 1970 have pushed the snow line uphill twelve hundred to fifteen hundred feet. Back east, a 2012 study found that about half of the 103 ski resorts in the Northeast will be forced to close by midcentury because of a lack of snow. A 2018 report by the snow advocacy group Protect Our Winters showed that high-snow years generate an extra $692.9 million at ski resorts and 11,800 extra jobs, compared with an average season. Low-snow years erase more than $1 billion in economic value and cost 17,400 jobs.

---

* As soon as 2030, the critical elevation for successful ski areas in the Alps will rise above five thousand feet. For reference, Chamonix, Davos, and Engelberg all have bases below four thousand feet; Garmisch and Kitzbühel are below three thousand. Of all ski resorts in Austria, 75 percent are situated below thirty-two hundred. (Skiing makes up 4.5 percent of the Austrian GNP.) Germany is even more vulnerable, losing 60 percent of snow-reliable resorts with just one degree Celsius of warming.

A man-made run near Corvara, Italy. Daniel Scott, climatologist for the University of Waterloo, found that keeping to Paris Agreement targets could prevent more than a hundred ski resorts in the US Northeast and eastern Canada from closing. If not, Dr. Scott says, seasons will be reduced by half, leaving just 30 percent of the resorts in the US Northeast capable of remaining open by midcentury.

I did not leave Zemp's office as excited as when I had arrived. I remembered the timeline Brad Markle had scratched into the snow and the 10,000-year climatic window we have been happily living in. It wasn't just that the cryosphere had fostered flora and fauna and created vast reservoirs of liquid fresh water with which humanity would prosper. Looking at Zemp's digital map of the earth, I saw how snow and ice had been freezing and melting for all time, sometimes quite quickly. Civilization was the variable. It had wrapped itself around the frozen world, assuming that it would never go away. The great cities of Europe were built on the shores of the Rhône, Po, Rhine, and Danube Rivers, all of which are fed by meltwater. The great metropolises and human populations of Asia sit along the Ganges, Yangtze, Syr Darya, Indus, Mekong, and Irrawaddy, all of which flow from the massive—and rapidly melting—glaciers of the Tibetan Plateau. Wherever you looked, frozen water has made our world.

The Little Ice Age

The perfectly calibrated engineering of alpine hospitality was on display the next morning at Rifugio Baita Cuz. A full breakfast buffet of croissants, Bavarian cream-filled doughnuts, scrambled eggs, charcuterie, yogurt, and a dozen local grains had arrived shrink-wrapped, boxed, and insulated with blankets on the back of a snowmobile late the night before—along with cases of wine, olive oil, vinegar, dried mushrooms, and *schüttelbrot* crispbread. A server from Sicily stood ready at the espresso machine while the manager and his wife set up the rifugio for the day.

Many of these winter traditions—food, dress, architecture, heating systems—harken back to another global warming event that brought year-round winter to the continent. A massive volcanic eruption on the Indonesian island of Lombok in 1257—in addition to solar activity, a shift in ocean currents, and a sudden growth of sea ice—triggered centuries of cooling in the Northern Hemisphere. The so-called Little Ice Age gripped midlatitude regions from sometime in the 1500s to the mid-nineteenth century, freezing the Thames River in London, Istanbul's Bosporus Strait, and even New York City's Upper Bay, allowing people to walk from lower Manhattan to Staten Island. Snow fell regularly in Europe, from Lisbon to Paris, sometimes all year, as in 1816, "the year without summer"—when there was widespread crop failure in Western Europe and a week of frost on the US East Coast during the last week of June.

In the seventeenth century, *frost fairs* were held regularly on the Thames. Participants would bowl, sled, dance, or ice skate and then visit—on the ice—a pub, the grocer, a barber, or any number of printing presses selling commemorative placards. As the notion of winter as the dominant season spread across the continent, populations adapted. The design and use of buttons, buttonholes, and linen long underwear became widespread, as did gloves, capes, and extended sleeves that sometimes draped to the ground. Churchgoers carried hot coals in steel spheres to keep warm during the service. Animal fur garments became so ubiquitous that Europe and Asia's supply gave out and the northern tier of America was colonized as a fur-trading empire. Five million felt hats, made from North American beaver pelts, were sold in England in 1688; between 1700 and 1770, England exported twenty-one million more. (England's population was only five and a half million at the time.)

Buildings adapted with the introduction of tiled stoves that could retain heat longer and with modern chimneys capable of spreading warmth safely through a house. Smaller rooms with corridors—which could be closed off for winter—replaced great halls and antechambers. Double-paned windows and insulated wall panels were also widely used. Artwork of the time by painters like Pieter Bruegel the Elder increasingly depicted novel winter activities like skating, sledding, and frost fairs. The snowflake itself became an icon—representing the Christmas season and the purity of Christ (Isaiah 1:18: "Though your sins be as scarlet, they shall be as white as snow").

Clean white aprons of snow covering dirty streets, factories, and livestock yards became a symbol of piety and resurrection. The legitimately creepy ghost of Saint Nicholas enforced this ideology around the Northern Hemisphere in various forms. In Switzerland it was Samichlaus who, with his black-clad henchman Schmutzli, traveled by donkey door-to-door with a giant book of sins—compiled by parents in the village—determining which kids got fruit and gingerbread and which would be stuffed into Schmutzli's sack, never to be seen again.* The king of winter in Slavic nations was Ded Moroz—Father Frost—who visited villagers with his granddaughter Snegurochka, who rode in a Russian troika fitted with sleigh runners. Farmers in Ukraine crafted *didukh,* Christmas decorations from the last sheaf of wheat reaped before the first snow. Christians in Alsace, France, adopted the pagan tradition of cutting down trees and decorating them indoors in the sixteenth century.† Toward the end of the winter season in Germany and Switzerland, Biikebrennen and Sechseläuten bonfires lit up the night sky to hasten the arrival of spring.

The cold spell also brought on catastrophic crop failure, famine, and disease. Lack of food in France, Norway, and Sweden at the end of the seventeenth century killed 10 percent of each country's population. Estonia and Finland lost 20 to 30 percent of their populations during the same time frame. Growing seasons contracted in northern Europe, causing grain prices to skyrocket and stoking bread riots and conflicts such as the Thirty Years' War. In the 1500s, the average height of Europeans dropped an inch because of malnourishment, and in Iceland, which was cut off from the world by miles of sea ice, the population fell by half.

The long, cold winters froze the Great Belt, across which Swedish soldiers attacked Copenhagen. French armies marched across the

---

* This gloomy, slightly terrifying narrative has persisted in the Christmas season, as seen in two made-for-TV stop-motion movies. There's the deeply dysfunctional and violent life of Rudolph and his outcast friends from the Island of Misfit Toys in *Rudolph the Red-Nosed Reindeer* and the prescient climate-change duel between the Snow Miser and the Heat Miser in *The Year Without a Santa Claus*. In the latter movie, the Heat Miser's mission is to melt every flake of snow on earth. (Tellingly, both these films were devised by former New York City advertising executives Arthur Rankin Jr. and Jules Bass to capitalize on the season and bring in good fourth-quarter returns.)

† The practice became so widespread that authorities tried to stop holiday deforestation in Freiburg in 1554 by banning it.

Netherlands' frozen rivers to attack the entrapped ships of the Dutch navy. Resulting political upheaval led to the rise of authoritarian regimes on a local and national level. Diseases caused by the freeze were blamed on Jews, while Catholics attributed them to the Reformation. The most violent decades of the Alpine witch hunts transpired during the Little Ice Age, as popes and bishops sought scapegoats for failing crops and political upheaval. Even in the New World, where explorers of the Golden Age of Discovery were setting up colonies, expeditions were hampered by historically cold winters that wiped out entire colonies in New England and Quebec. To illustrate the power of climate change, the great winter that swept across

One of the earliest Roman roads through the Alps, Via Claudia Augusta, was engineered in 15 BC to connect the Po River near Venice with what was then called Rhaetia (now southern Germany). The road stretched nearly five hundred miles, passing through Reschen Pass, and became the primary thoroughfare connecting the northern and southern provinces of the Roman Empire.

Europe was triggered by a temperature shift of just one degree Celsius in most regions.

A gust of icy wind accompanied two frosted touring skiers who walked into the rifugio from the storm. They must have started before dawn. The wind outside was stronger than yesterday; it looked like a whiteout though the window. The forecast called for gusts of 140 miles per hour that day. All the lifts were closed again.

Fully caffeinated and geared up, we put on our skis and started down from the rifugio. It was eighteen degrees Fahrenheit, with a mild thirty-mile-per-hour wind. We followed Cominetti to the base area, where he ushered us away from another closed lift and into a taxi headed for Passo Fedaia. We dropped off our gear at Rifugio Capanna Bill, where we would spend the night, then skinned up the pass alongside a narrow road. Tourists drove by slowly, snapping pictures of us from their cars. The drifts were icy and tricky to navigate, so I shouldered my skis and walked on the pavement instead. The terrain was similar to the day before, as was the ghost-town quality, though the pitches were longer and steeper and, somehow, icier. Even with the taxi ride, we were only a third of the way around the circuit—with three valleys yet to ski. If the wind didn't die down, there was no way we would make it. Even Cominetti started to show signs of fatigue as we started up another headwall.

Higher up, sections of an intermediate groomer induced an honest-to-God sense of terror with each step. I knew it was serious when Cominetti froze midstride for a full minute, then took off his skis carefully and chipped little steps into the slope, moving sideways to get to softer snow. I tried to follow, kneeling, clawing, and eventually strapping my skis to my backpack and walking up the edge of the trail.

We had climbed two thousand vertical feet in two hours. The last meal I ate was breakfast; it was now three in the afternoon. I wasn't sure what we were trying to ski, or if I would make it to the top. I'd been so focused on not falling that I hadn't looked behind me the entire climb. When I finally did, at a switchback in the trail, I saw the giant maw of the Marmolada Glacier directly across the valley.

It is hard to explain the look of a mountain glacier like that, the sheer face, ancient gray ice clinging to the summit, cracked and bunched up in the middle, pure white on top. I imagined what the doges of Venice or the crusaders thought as they rode by.* Seventy years ago, the glacier extended to 7,800 feet. Cominetti pointed out one crevasse that had been a 250-foot-deep people-eater during World War I but was now barely a divot in the snow.

It was impossible to imagine the Marmolada gone in fifty years. And yet it was happening. Cominetti was too preoccupied navigating an icy headwall to take in the view. All of us now had our skis on our shoulders as we made our way up. The gusts were so powerful it was difficult to stand. Near the top, I hid behind a large concrete balustrade attached to the summit station while the group tried to reach the couloir. It didn't happen. The wind had scoured all the snow, and the couloir was rock-hard. Cominetti didn't mind. He had a different plan—perhaps all along. He was ditching us to ski the opposite side of the mountain—where, by chance, his house was located, as well as his daughter, who was visiting for the night. "See you tomorrow!" he yelled as he poled across the summit plateau and vanished down the trail.

* In the 1400s, Leonardo da Vinci climbed Monte Rosa's flanks, 150 miles west of the Marmolada, to study glacial movement. A royal mathematician, Donato Rossetti, studied snow crystals in the late 1600s and published *La Figura Della Neve* (The shape of snow). The oldest snowfall records on the planet were kept at a meteorological observatory in Turin beginning in 1675. A recent analysis of the 345-year series confirms a steady downward trend, especially after 1890.

12

The Ice City

The wind died down sometime during the second night of our tour, and on our third morning on the Sellaronda, like a magician's trick, the lifts reopened. Cominetti showed up on time, somewhat frazzled that we were so far behind on our itinerary. Halfway through the trip, we had made it only a third of the way around the circuit. We still had several valleys and a dozen lifts and trails to reach in the next thirty-six hours. We packed our bags and skied to the Marmolada tram, then rode to Punta Rocca, at 10,856 feet.

The summit station was more like a mini mall with several floors, the top one being an observation deck from which you could see most of the Dolomites. The range looked like a carpet of granitic thumbtacks pointy side up. Mount Tofane (10,5643 feet) on one side and Civetta on the other. The mountains did not rise from a common ridge but were each great massifs unto themselves. The range was once a chunk of a tropical seafloor before the European and African tectonic plates collided. Sheer dolomite walls riddled with fossilized sea creatures were yellow-pink, with striations of snow running horizontally on ledges and rifts. Where a snowfield had enough surface area to form, the sun glanced off the thin, icy crust, sending flashes of golden light across the range.

Punta Rocca holds a place in history for one of the more bizarre roles ice and snow have played in world history. In World War I, Italian and Austro-Hungarian soldiers were stationed along the Austro-Hungarian Empire's southern border when Italy declared war in 1915. Conditions were miserable as troops fought at elevations of up to twelve thousand feet—where oxygen levels are 30 percent less than at sea level and temperatures sometimes dipped to minus twenty degrees Fahrenheit. One afternoon, a lieutenant named Leo Handl dug a tunnel beneath the Marmolada to reach the top of the glacier without being picked off by Italian snipers. Similar to the previously mentioned scientists Johannes Georgi and Ernst Sorge in Greenland—and maybe even Christine Foreman's glacial bacteria—Handl and his compatriots found a quiet, safe, and surprisingly warm refuge inside the ice. Soldiers eventually dug more than ten miles of tunnels, caverns, dining halls, first aid stations, generator rooms, and bunk rooms, forming an ice city beneath the Marmolada. Tunnels were barely head-high and were arched in the middle to allow easier passage. Some passed through ice, then rock, to reach the far side of a ridge or summit. Some were two hundred feet beneath the surface. Cable cars were strung to the edge of the glacier, just beyond the field of fire, where soldiers carrying supplies would disappear into the tunnels.

We took the summit station elevator three floors down to the Marmolada and joined a couple hundred other skiers getting ready to push off. To the left, the upper snowfields of the glacier undulated down to a gray granite ridge. To the right, a slope ran toward the rifugio where we had stayed the night before. In the middle was Sasso Di Mezzodi peak and Val Cordevole leading toward Belluno. Still cautious with my knee, I skied on the piste as Cominetti and the group branched off into the backcountry. Lustenberger and Monego cut graceful lines through stranded aprons of powder between arêtes and boulders, white clouds of snow swirling behind them. Reddick moved down the hill like an ermine, creeping along rock outcroppings, hiding behind snowdrifts, finding new angles to photograph them from.

I took my time skiing down, making careful, even turns, envisioning the mass of ice beneath the slope, the ice barracks and meeting rooms, all strung with electric lighting and stacked with rifles and explosives. The horrors of climate change and the end of winter will surpass those of World War I. Not all at once, but gradually. If emissions continue at current levels, the world's mortality rate is estimated to increase seventy-three deaths per hundred thousand by 2100—the same amount killed every year by infectious diseases, including tuberculosis, HIV/AIDS, malaria, and dengue and yellow fever combined.*

Most of the data that Zemp and others used to make these dire projections comes not from in situ measurements but from an expanding fleet of satellites pointed at the cryosphere. A few weeks before the trip, I got to sit in on a conference for a group of rather nerdy souls who manage this bird's-eye view of the cryosphere. It was at the University of Bern, a mile from the history museum I had visited. Chairman Stefan Winderle opened the ninth Workshop on Remote Sensing of Land Ice and Snow with a statement about where snow definitely was not. "It rained to twenty-five hundred meters last night," he said in a thick German-Swiss accent. "It is January, and it feels like spring when you get off the train, yes?" Some of the scientists nodded, but most did not seem amused by his observation.

* Data from an ice core recently retrieved from the Alps suggests that an abnormal cold spell from 1914 to 1919 significantly increased fatalities during World War I. (One December 1916 storm dropped thirty feet of snow, creating avalanches that killed ten thousand soldiers along the Italian front in a matter of days.) Part of the cold snap was due to the war itself as an influx of dust and particulate from explosives blocked the sun. In another macabre twist, cooling kept mallard ducks—primary carriers of the 1918 Spanish influenza virus that would kill fifty million people—in Western Europe. Instead of migrating to Russia as they normally did, they remained, probably increasing the spread of the virus there.

The United States had just experienced its warmest December and January in recorded history. The snow line in Bern had risen from sixteen hundred feet to thirty-three hundred feet in a few decades. The scientists at the conference, however, were more concerned about the process of data collection, the data's accuracy, and how to get better information.

Fluorescent bulbs hummed above ten rows of faux-wood desks. About a quarter of the room was filled, making the place feel awkwardly cavernous, like a party where only a few guests showed up. On either side of the classroom, posters with titles like "Accuracy Study of Snow Cover Maps Based on AVHRR [advanced very-high-resolution radiometer] Data with Different Spatial Resolution" and "Light Absorbing Impurities in the Olivares Catchment, Central Chile" displayed pie charts and regional images deciphered from Envisat and other earth-observing satellites and aircraft. Represented here were the massively complex and fruitful Global Cryosphere Watch and Global Climate Observing System, both run by the World Meteorological Organization.

Measurement is the key to solving climate change. Many issues discussed at the conference revolved around how it wasn't working. There was no way to measure soil evaporation, so don't try. There was also no way to measure permafrost thaw or high-latitude snowpack ablation before melt season—that is, a midwinter drought—so don't try that, either.* One conversation that dragged on for almost forty minutes concerned what to do with pixels in satellite imagery that were half white and half dark. How do you measure those? Can you toss them out? What if the dark spot is black ice? What about shore-fast ice that might appear white but is definitely

* The recently launched $1 billion NASA ICESat-2 satellite might be able to capture thermokarst features and changes in topography, essentially measuring some permafrost thaw. Elevation accuracy on ICESat-2 is about the thickness of a pencil.

not real ice? "The images can't be read like that," said a Russian scientist sitting in the otherwise-empty front row. "It's misleading to say that they can. There is a gap there, and we don't know how to do it."

This was the slow churn of science trying to catch up with the rapid pace of climate change. One pixel will not change our future, but a hundred trillion pixels could very well influence the course of international climate change treaties, domestic governmental regulation, trade agreements, and the timetable of the Great Melt.* By measuring things like the ratio of dark earth to snow-covered earth from space and by monitoring snow and ice extent and the length of winter, the thickness of glaciers, and the depth of snowpack, scientists can piece together a global picture—similar to Zemp's—with far greater specificity. Like the fact that one metric ton of carbon dioxide emitted into the atmosphere will melt three square meters of sea ice. Or this cute, self-effacing stat from Winderle: it took most conference participants an average of eighteen tons of carbon dioxide to travel to the conference. So each attendee was responsible for melting fifty square meters of sea ice. And another: "What we are monitoring here, we will lose in the near future. So everyone here is about to be out of a job." Or the fact that many white pixels in the mountains represented the only source of drinking water for two billion people.

The coming water crisis was laid out in a 2015 study titled "The Potential for Snow to Supply Human Water Demand in the Present and Future." Published in *Environmental Research Letters,* the paper identified 421 major snowmelt catchment basins in the

---

* A recent study of earth observation satellite data from 1994 to 2017 by researchers at the University of Leeds in the UK revealed a massive acceleration in the rate of worldwide ice melt, putting the cryosphere's demise on the IPCC's "worst-case scenario" track.

Northern Hemisphere. Of them, 97 were "exposed to a 67 percent risk of decreased snow supply this coming century." These figures mean that there is a 67 percent chance that water stored in snow will run out. Of the most endangered catchments, 7 are in the US West, including basins in Sacramento, Coastal California, San Joaquin County, and the Rio Grande. Up to 75 percent of the water used by farms and cities in the American West comes from snowmelt. The Colorado River alone, which is filled primarily with snowmelt from the Rocky Mountains, is losing almost 10 percent of its flow with every increase of 1.8 degrees Fahrenheit.* The Colorado supplies fresh water for forty million people, including those in the cities of Los Angeles, San Diego, Phoenix, Tucson, Denver, Salt Lake City, and Albuquerque. The greatest threat in Europe exists in the Ebro-Duero catchment in Spain, home to thirty-two million people, and the South Apennines in southern Italy, home to one million.

Just seventy-eight glaciated water towers around the world provide fresh water for two billion people downstream—from Peru to China to the Rockies and the Alps. And they are all melting. If we somehow keep warming to 1.5 degrees Celsius, temperatures in the Himalayas will hit 2.1 degrees Celsius. Rivers like the Syr Darya, which flows from the snowy Tian Shan Mountains in Kyrgyzstan and eastern Uzbekistan, are what worry scientists and the twenty million people living along it most. Monsoon rains will likely offset the lack of glacial outflow in most South Asian regions. But the Syr Darya lies outside the monsoon system and has nothing to replace meltwater once it stops flowing. The Indus Basin, fed by the Himalayan, Karakoram, Hindu-Kush, and Ladakh ranges, faces a similar problem—putting two hundred million people in Afghanistan, China, India, and Pakistan at risk of water scarcity. Even in mon-

* In a disturbing trend, Wall Street investors now purchase water rights along the Colorado River, upending municipal control by selling them to the highest bidder.

soon country, the lack of glacial meltwater is already drying up farmland, pastures, and water districts. In 2019, a weak monsoon led Chennai, one of the largest cities in India, to nearly run out of water as glacial runoff stopped.

I had a plane to catch and left the conference after the lunch break. Outside, the hottest January in recorded history had just been logged. It was fifty degrees Fahrenheit and raining. I walked past a thousand bicycles chained up outside the central train station. More were chained to street signs, steel racks, and an indoor, robotic storage system from which a cyclist could recall a bike by entering a code. There was hardly a car on the street. It was quiet, no horns, not even yelling. A fully enclosed two-seater recumbent bicycle parked nearby had solar panels on the roof, locking doors, and a stereo-trip computer console between the seats. The science is clear; the math has been calculated and confirmed. Our best chance of survival is to change the way we create and consume energy immediately. It was happening here; I wondered why it was lagging so far behind in the United States. Looking at a young family of four carrying their skis into the train station to catch a ride to the mountains, a zero-carbon future looked like a pretty good idea.

The Clockworks

Since we were too far behind on the circuit to catch up in a day, Cominetti hailed us another cab at the bottom of the Marmolada, and we headed to the final leg of the Sellaronda—the Cinque Torri formation in the Nuvolao Group.* At the top of the chairlift, which threads through dolomite rock formations in an uncannily Disney-

* Do not confuse this with the Disney-esque Cinque Terre, where honeymooners and semester-abroad nomads trek between quaint seaside villages on the Italian Riviera. There are similarities, however.

ish way, the five pale gray towers of the Cinque Torri jutted from the summit like aged, stubby fingers.

Golden afternoon light touched the spires. Depending what direction we skied, we could continue to the Sellaronda by heading toward Lagazuoi, Col Gallina, or Falzarego Pass. The lifts were closing soon, so we skied down to the base and caught a ride to our final stop on the trip, Rifugio Passo Giau, set on the pass it was named for. The rifugio was built in the shadow of Cima Ra Gusela, a massive rock tower surrounded by alpine fields. We walked through a carved wooden gate and down a stone path to a sundeck. A wood carving of an owl hung over the front door, as did six engraved red squirrels of the renowned Scoiattoli di Cortina (the Cortina Squirrels), a climbing club that was founded in 1939.

Inside the rifugio, wooden rafters ran between whitewashed walls above hand-carved doors numbered one through ten. An old Singer sewing machine sat on a table near the stairwell. At the top of the stairs, three motorcycles—Zündapp, Moto Parilla, Moto Guzzi—sat on a glass panel directly above the cash register.

Igor Valet welcomed us at the door and showed us to our rooms. Igor's parents built the rifugio in the 1990s, on the site of an old roadhouse that had sheltered passersby since the 1930s. The place was still a family operation. Igor was in charge of maintenance and personnel. His father was the head cook; his mother the head food server. The place was busier in the summer, when auto tours and cyclists stopped by for a bite or to spend the night. Winters were quieter, Igor said, especially since recent winters started a month later and ended a month earlier.

Igor, whose calloused hands were the size of oven mitts, was a craftsman-philosopher. He never looked at screens or computers, he said. He preferred the works of Hermann Hesse, saying that our only salvation from ourselves and the problems we have created—like climate change—would come from the stars. "When you stand on the deck at night and see the stars," he said, "it is like they are right there. There is nothing between you and them."

His shop in the basement was a clockwork of his life as a caretaker. He was in charge of plumbing, heating, structural, and any other repairs the rifugio needed. There was a band saw and a welder in the corner, a drill press, two pegboard walls layered with tools, and every gadget he needed to keep guests warm, fed, and safe at seven thousand feet in the Dolomites. "We can have times during a storm when no one can get in or out of here for a week or more," Igor said. If the water lines burst, it was his responsibility to get them flowing again. He was so invested in this task, and so keen to have customers understand the intricate web of challenges hoteliers in the Alps faced, that he printed photos and a description of the cistern and water system at the bottom of the lunch menu and a snowfall chart on the back. "I want people to know that water isn't free," he told me, smiling. His prized tool: a lime green Ferrari snowblower, with shiny galvanized chains wrapped around each tire. "I like to tell the girls that I drive Ferrari," he said.

The rifugio maintains a supply of food and plenty of wine. "In case the road closes," Igor's mother, Claudia, told me that night in the bar. Igor poured us a red Rhône blend produced in Napa Valley, of all places, made by longtime customer, Gary Erickson, the founder of the Clif Bar & Company. Claudia flipped through a book about the Scoiattoli di Cortina as we sipped, pointing out people every few pages: "There is my father. There is my cousin." A black-and-white image behind our booth showed her father climbing an overhanging rock wall, with ropes and wire Prusiks hanging from his waist. Another showed him wearing long knickers, wool socks, and a Scoiattoli di Cortina sweater, practicing knots with a friend.

That night, Claudia seated us across from a doctor and his wife from Tuscany, who had been coming to Rifugio Passo Giau every

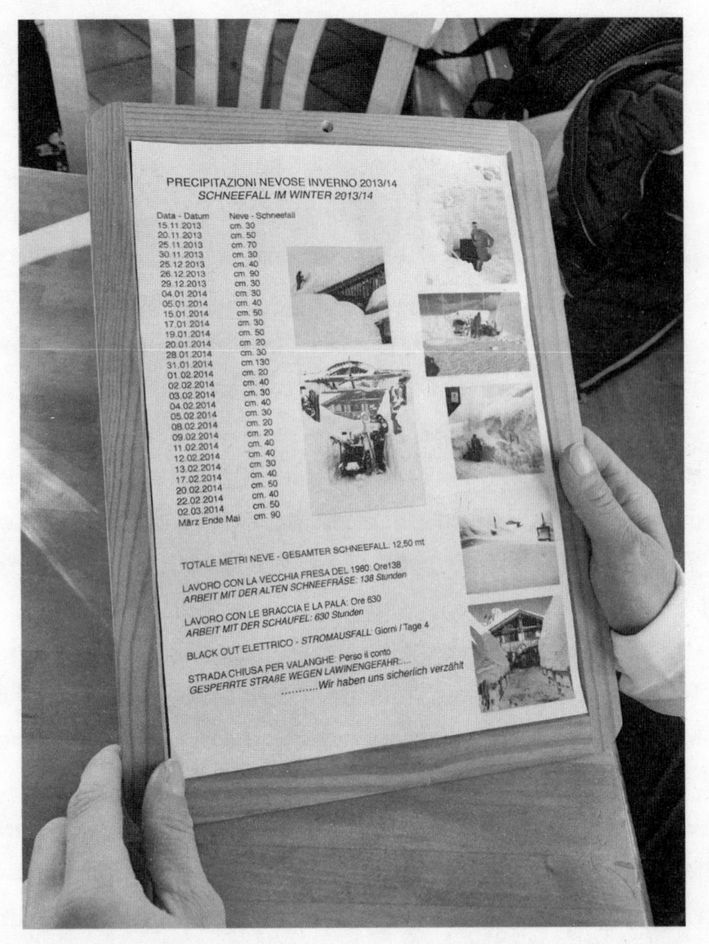

PRECIPITAZIONI NEVOSE INVERNO 2013/14
SCHNEEFALL IM WINTER 2013/14

Data - Datum	Neve - Schneefall
15.11.2013	cm. 30
20.11.2013	cm. 50
25.11.2013	cm. 70
30.11.2013	cm. 30
25.12.2013	cm. 40
26.12.2013	cm. 90
29.12.2013	cm. 30
04.01.2014	cm. 30
05.01.2014	cm. 40
15.01.2014	cm. 50
17.01.2014	cm. 30
19.01.2014	cm. 50
20.01.2014	cm. 20
28.01.2014	cm. 30
31.01.2014	cm.130
01.02.2014	cm. 20
02.02.2014	cm. 40
03.02.2014	cm. 30
04.02.2014	cm. 40
05.02.2014	cm. 30
08.02.2014	cm. 20
09.02.2014	cm. 20
11.02.2014	cm. 40
12.02.2014	cm. 40
13.02.2014	cm. 30
17.02.2014	cm. 40
20.02.2014	cm. 50
22.02.2014	cm. 40
02.03.2014	cm. 50
März Ende Mai	cm. 90

TOTALE METRI NEVE - GESAMTER SCHNEEFALL: 12,50 mt

LAVORO CON LA VECCHIA FRESA DEL 1980: Ore138
ARBEIT MIT DER ALTEN SCHNEEFRÄSE: 138 Stunden

LAVORO CON LE BRACCIA E LA PALA: Ore 630
ARBEIT MIT DER SCHAUFEL: 630 Stunden

BLACK OUT ELETTRICO - *STROMAUSFALL*: Giorni / Tage 4

STRADA CHIUSA PER VALANGHE: Perso il conto
GESPERRTE STRAßE WEGEN LAWINENGEFAHR:...
...........Wir haben uns sicherlich verzählt

Igor Valet's snow chart, located on the back of the Rifugio Passo Giau menu.

winter for twenty-five years. Dinner was a local specialty, beetroot ravioli cut and pressed into plump little ears by Claudia that afternoon. A woodstove in the corner heated a large ceramic chamber that then heated the dining room. Each table was set with linen tablecloths and wooden trays. The ceiling was handmade wood panels, and the ceramic chandeliers held electric lightbulbs. Out the window, you could see the five fingers of the Cinque Torri.

The Valets joined their friends after dinner and launched into stories from the old days. The doctor had brought Tuscan *cantucci,*

Claudia Valet showing photos of her father, one of Cortina's first mountain guides. More than 670 million people live in what the United Nations defines as high mountain regions. *The number is expected to grow to 840 million by 2050. All will be affected by rising temperatures, extreme weather events, permafrost thaw, avalanche danger, drought, and crop failure in the coming century.*

like biscotti, and Igor fetched traditional *vin santo,* sweet wine, to dip it in. When we finally headed to our rooms, the group, now with Igor's father, Diego, who was still wearing his apron, roared with laughter as Igor delivered more bottles of wine.

I conjured an image from earlier in the day as I fell asleep in my room. I had been looking out the front door of the rifugio. The sun was going down, creating a yellow sky above the yellow dolomite rock. Some of the higher snowfields facing west turned hot pink; the snow in the shadows darkened to a dull shade of blue. Route 638, which connects to Cortina d'Ampezzo, wound down the valley, a black strip bordered by five-foot snowbanks. There was no traffic, just the fading light and tiers of snow, rock, sky, and cloud.

One lone skier stood on top of the pass, an ink black silhouette hunched over to click into his bindings. As the last alpenglow left the top of the five towers, he pushed off and arced down the slope. Snow sailed behind him and hung in the air like a translucent veil. He made a few more turns, then angled toward the parking lot, where he slid to a stop on a large snowdrift and sidestepped to the pavement. Then he took his skis off, tossed them in the back of his car, and headed down the pass toward home.

13

The Lost and Found Memories Office

Cominetti did not join us the last day. His father was ill, and he had to drive to Genoa to check on him. Monego took Reddick and Lustenberger to hike a 2,000-foot couloir and showed me the best route to head down so that I could finish the circuit and record one last interview. The trail was sublime—a steep, rolling groomer that was easy on my knee and lasted for miles. A serrated rim of yellow rock 2,000 feet tall wrapped around the right edge of the run. Stands of old-growth Swiss pine appeared halfway down, growing in steep draws. The route was once used by the Alpini to resupply advanced positions during World War I, and I spotted some of the fortifications and tunnel entrances as I skied by.

The trail ended at a snowy parking lot, where I caught a taxi to Corvara, completing the Sellaronda at the white columns of Hotel La Perla, where we had begun. One of the last living pioneers of winter waited for me inside. Ernesto Costa was a descendant of original Ladin settlers. When he was a boy—he is ninety now—there was hardly any winter economy in the Dolomites, just a modest timber trade and a few traditional Ladin crafts like silver filigree and homemade wooden toys.*

* Wood harvested from the Ladin valleys was used in the lucrative Venetian trade of building boats for wild-eyed crusaders in the Middle Ages. Ladins also used wood from their forests in the 1800s to shape skis for Napoléon's army.

With no winter income, men often had to leave the valleys for months or years at a time to provide for their families. The farmhouse Costa grew up in across the street had no heat or indoor plumbing. This was the 1930s, not long after Gabriele D'Annunzio's troops revolted in Fiume and socialist-journalist Benito Mussolini overtook Milan with his *fasci di combattimento*. As a young boy, Costa went to school for a few years, then apprenticed as a plumber. By the time he was a teenager, he realized that he and his family were not making a living; they were merely surviving.

A boy in town and a friend of Costa's had taught himself to ski using Matthias Zdarsky's manual—then in its seventeenth printing—and got a job as a ski instructor at a one-run ski hill in Corvara. The boy made more money in one winter than either he or his friend had ever seen. Costa soon found an old pair of skis in the village and learned the basics. His friend got him a job teaching, and when construction of the Col Alt chairlift began, Costa used his plumbing skills to help disassemble World War II tanks to use in the tower foundations. Ten years later, he'd made enough money teaching to open La Perla.

Costa met me in one of La Perla's four dining rooms. The hotel is now one of the finest in the Alps, with Michelin star ratings, the highest-rated wine cellar in all of Italy, and a roster of high-profile guests. He wore a navy blue cashmere V-neck over a blue-and-white striped oxford. His prodigious eyebrows arched, and crow's feet bunched around his eyes as he smiled and greeted me. Our translator was a young woman named Stephanie, who wore a La Perla uniform: hair twisted in a bun, traditional black Ladin dress, and a pink apron. She sat on my left, Costa on my right, creating an awkward conversational triangle in which I was wedged, mostly silent, in the middle while the two chatted in Ladin.

An air of importance enshrouded Costa, perhaps because he had created so much of the winter culture that I'd lived in most of my life. The Dolomites were barren in the winter before he and his peers arrived. Afterward, mountains across Italy buzzed with thousands—millions, over the years—of euphoric guests experiencing the majesty of the winter season. Costa was the final link in the Alpine evolution I'd come to see—an evolution that began with the Rhaetians and ended with the worldwide phenomenon of winter tourism. My question for him: what would happen to it when the snow was gone?

Costa didn't answer at first. Three espressos arrived at the table, and he fired off stories about the old days to Stephanie, who giggled and nodded and occasionally narrowed her eyes and scolded the old man. The two had such rapport that Stephanie often forgot to translate the conversation—and at one point seemed to forget that I was sitting between her and her boss. Morning light pressed through a small window behind him. The scent of melting butter and rosemary escaped the kitchen. The ceiling rafters were hand-carved, and the tablecloths, metalwork, embroidery, furniture, and pretty much everything else down to the pillows stacked in the corner of the banquette were all handmade in Ladin tradition.

During a break in the conversation, I posed the question again, and Costa replied, "There have always been changes. Thirty years ago, we saw there was less snow, so we bought snowmakers. Another time, I remember the Marmolada was so filled in with snow that I skied it all summer. I also remember the winter of 1989, when I picked wildflowers on the slopes above the hotel in February... The snow will come again."

Costa, an old-timer in every way, did not seem to grasp the immediacy and inevitability of climate change. Like many older skiers I had interviewed in the Alps, he did not believe in *anthropogenic* climate change. Temperature swings were natural, they said. Someday it would be cold and snowy again. Residents of the Alps had a unique perspective on this issue; their families had lived there for so long, they had seen and documented climate change for dozens of generations. During the Little Ice Age, the Dark Ages Cold Period, and the tail end of the Medieval Warm Period, entire Alpine villages were bulldozed by advancing glaciers, rebuilt when they receded, then razed again. Valuable grazing pasture used by farmers in the summer—there is a very cool word for this seasonal shifting of livestock: *transhumance*—was covered and uncovered by glaciers so many times that a system of laws was enacted, governing who owned land beneath the ice when it receded and who was liable if a glacier knocked over a building. When the Mer de Glace Glacier near Chamonix, France, encroached on the village of Les Bois in the 1700s, villagers brought in a bishop, marched to the tongue of the ice, sang hymns, and prayed. (The ice had already taken out the villages of Bonnenuit, Le Châtelard, and La Bonneville.) Bishops often sprinkled holy water on the ice to slow it, and villagers erected statues of Saint Ignatius in its path to keep it from coming back.*

The difference between then and now is that those changes were sparked by natural causes and typically took place over centuries, allowing time for people to react. Present-day anthropogenic warming will change our world in decades, making it almost impossible to adapt the Daedalian web of infrastructure we have built around the cryosphere and threatening hundreds of millions who live there, and billions more downstream.

I had to pack and get ready for an early-morning flight home, so after our interview I said goodbye and hopped in another cab. We drove thirty minutes down a steep, winding road lined with whitewashed churches, larch-sided barns, snow-covered pastures, and a

---

* Ancient settlements in Europe saw something similar. Humans wandered north from the Pyrenees, Jericho, Turkey, and the Sahara at the end of the last Ice Age, only to be hit with a drop of eighteen degrees Fahrenheit during the Younger Dryas period, which froze the planet for nearly fifteen hundred years. Glaciers started growing again, locking in fresh water and causing widespread drought. Caught off guard, humans devised a new way to survive: farming. A few thousand years later, another mini Ice Age forced new farming communities in the Fertile Crescent to change yet again—pooling resources to form the world's first cities and nations.

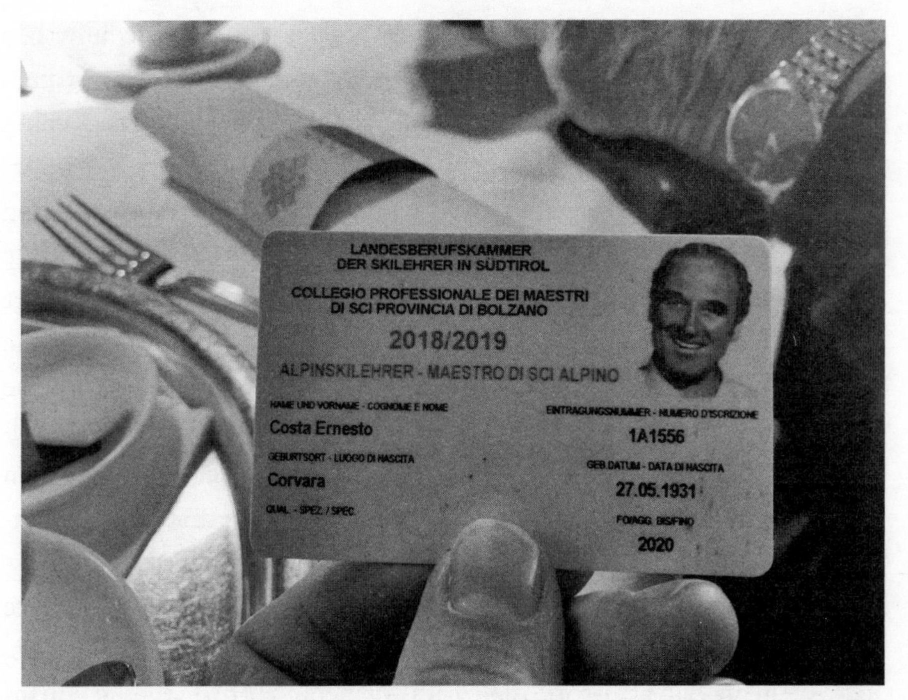

Consider the brevity of human time. Ernesto Costa began teaching skiing seventy years ago. It has been just ten very long human lifetimes, or forty generations, since people of the Middle Ages searched for dragons and Devil's pulpits in the Alps. The first travelers breached the high Alps just six thousand years ago, or sixty human lifetimes. And yet, it has taken us just five generations to put more carbon dioxide into the atmosphere than has been seen in three million years.

seemingly endless procession of hotels and restaurants. The amount of development in the valley was astounding, and I couldn't help but notice the avalanche fields and frighteningly tall towers of rock hanging above it.

A conversation that I'd had with Michael Fässler, curator of the Swiss Alpine Museum in Bern, had shed some light on how the people in the Alps would react to the end of winter. He'd alluded to the connection Europeans have cultivated with the mountains, snow, and winter and how climate change there was not just an economic or a societal crisis, but a crisis of identity. "As a kid in Switzerland,"

Fässler said, "you learn how to swim, how not to drown, how to ride your bike and how to ski; that's just what you do. You go skiing with your school class for a week once a year. You go on the weekends. You watch famous ski racers on television at home. But now skiing, not only as a sport but also as a lifestyle, is vanishing."

Fässler met me in a chic café inside the black minimalist building that houses the museum. The café was packed and, like many things in Switzerland, seemed to be an example of the way museums ought to operate. Festive. Cool. Excellent coffee. "People come here for the café," Fässler whispered. "They don't even go to the exhibits!"

Fässler's job was to preserve things, and with the end of winter in sight, he and his staff had been preserving a lifestyle that most in the Alps thought would never change. "Should winter sports be gone due to climate change in fifty years, we asked ourselves, 'What should we leave our successors in our collection?' " he said. "Not just the gear, but a history of stories, of personal history—we just want to know what it meant to people."

The exhibit, titled *The Lost and Found Memories Office,* was taking shape in a basement room the day I visited. Fässler led me down a flight of concrete stairs to a taped-off glass door. Inside, a row of old T-bars hung in front of a wall of skis, including short, fat, snowshoe-style wooden boards and fiberglass skis of local Olympic gold medalists. Next to them were a pink-and-red one-piece snowsuit, a rack of replaceable aluminum ski tips—used in the old days when a tip broke—and a wall of blue-and-white ski resort trail maps with hundreds of curving white lines representing trails where entire generations of children had learned to ski, travel in the mountains, and immerse themselves in winter.

As a new parent, Fässler was also feeling the same protective instinct that I had: you can't take winter away from our children. On top of the worldwide cataclysm the Great Melt would bring, you can't take away skiing, pond hockey, sledding, ice fishing, snowball fights, and thawing

Skiing was so much cooler back then. (Photo courtesy of the Swiss Alpine Museum, Bern, Switzerland)

out in front of a fire with a mug of hot chocolate. The exhibit was eerie, like a catacomb of childhood memories: falling off T-bars, freezing your tongue to a chairlift, skiing in the rain wearing trash bags, snowball fights, Christmas blizzards. I wondered if winter would become mud season by the time Grey reached my age. I wondered what would replace the tangle of eleven hundred ski resorts that now occupy the Alps.

"Children born today will never see winters like we did," Fässler said. He recently performed an experiment with a group of kids to test his theory. He showed them a thousand old photos of winter and skiing in the Alps. Most, he said, couldn't believe how much snow there used to be. "They asked, 'Is that really Switzerland?' and that was a bit shocking for me," he said. "Winter is getting more and more like an abstract concept, like a fantasy world portrayed on television and in magazine ads."

Back in the cab, a few miles from La Perla, an image of the future materialized out the window. A small resort had made snow for a single trail. The entire mountain was green except for a thin strip of snow running a thousand feet down the right side. Snowmakers lined the run; grass grew at the edges. A procession of skiers etched turns down the artificial surface. It was a sad scene: the grandeur of the Alps reduced to a carnival ride. But it was more than that, I realized. Imperiled skiing, winter culture, and tourism jobs were merely canaries in the coal mine. The real catastrophe had just begun. As the frozen water towers of Europe vanished, how would the continent's matrix of borders, electrical grids, breakwaters, shipping canals, vineyards, farms, and general infrastructure hold up? Hydropower is bound to nosedive right when renewable energy is most needed; heat waves will cook lowlands no longer watered by cool glacial runoff; crumbling mountains and glaciers will annihilate villages and cut off valleys, tunnels, and vital trade routes. As I watched a skier carve tiny turns down the strip of snow, the workings of the Old World looked precarious as climate change gained momentum.

If Europe was the case study I had come to see—and one of the wealthiest, most developed regions on earth—challenges for the developing world were daunting. I couldn't imagine what would happen in places like Nepal, India, Kazakhstan, and to the billions of people who had unknowingly wrapped themselves around the outflow of the cryosphere. Just think about the four hundred million people living in the Ganges River basin, another four hundred million on the Yangtze. Where will everyone go when the rivers dry up? As I packed my bags that night for the journey home, it dawned on me that you don't have to live in the mountains to feel the wrath of winter's end; it will find you.

White Earth

14

A Bad Omen

On the edge of town, past the schoolhouse, the supply boat landing, and the speedboats sitting high and dry on sea ice, twenty-eight houses that make up the village of Kulusuk, in Greenland, stood with their gables angled northeast. They were built in traditional Danish design, a hand-me-down from their onetime colonial overlords who shipped the first such buildings across the Norwegian Sea with Lutheran missionaries in the 1700s. The colors once indicated the function of the houses: magistrate (black); fish processing plant (blue); hospital (yellow). Three windows with white trim were positioned on the end of each house—one near the ridge, two set symmetrically below it. Looking out from those windows, you would see what will one day be the last stronghold of winter: ripples of ice and rock, a frozen sea cross-hatched by ridges and drifts, and so much snow that the Inuit people who first settled here millennia ago named the island Kalaallit Nunaat, or "White Earth."

Inuits arrived in Greenland around five thousand years ago and had since explored nearly all of its 736,000 square miles—95 percent of which is ice. A turf hut, similar to the first ones built in Greenland, still sits on the outskirts of town. The stone-and-grass home has a small, arched doorway and no windows. The roof, which is now long gone, would have been supported by whale ribs and driftwood and covered with caribou and musk ox skins. Families back then burned seal and whale blubber in stone chalices for light and often slept in a pile, keeping each other warm beneath layers of skins and furs. (Minus sixty degrees Fahrenheit is not uncommon in the winter.) Like present-day homes, doors on the turf huts were angled away from the northeast, because for all time, even before humans arrived on the world's largest island, that was where bad weather came from.

Around 95 percent of Greenland sits north of the Arctic Circle. Weather there is a pure, unobstructed force that has affected civilizations and climates far beyond its shores. The wind doesn't build there; it arrives like a punch, pounding buildings and people with hail and snow. Gale winds of 70 miles an hour are frequent. (The fifth-fastest wind gust in recorded history was logged at the Thule Air Base in northwest Greenland: 207 miles an hour.) The wind cuts through deep valleys and ocean inlets, joins katabatic winds rushing down 10,000-foot mountains, then rakes through town, knocking over steel shipping containers and tearing up anything not tethered to the ground. Winter starts in September and lasts until June. Average winter temperatures hover around twenty-five degrees below zero Fahrenheit, preserving the world's second-largest body of ice.

Three narrow roads bisect Kulusuk, each lined with streetlights affixed to fortified steel poles. The roads are not like ones you know; they are pathways of packed snow where ATVs, snowmobiles, and hand-hewn dogsleds hauled by Greenlandic huskies rumble and slide. They are also surprisingly full for a small town, as villagers run errands and haul harp seals, halibut, shark, and polar bears back from the wilderness to be butchered. Kids walk to and from a bright red schoolhouse—once the (blue) Royal Greenland fish processing plant—and meet up outside the town's only store to gossip, drink Pepsi, and sled down the hills.

When I flew into Kulusuk earlier that day, the village looked like a handful of multicolored Lego bricks dropped onto a rolling promontory of ice, snow, and black volcanic rock. Two hundred people live there. Only sixty thousand, 85 percent of whom are Inuit, inhabit the entire island. A sea of shattered, moving ice braced the peninsula on all sides. To the north and west was a latticework of mountain ranges, each one massive and stark against the flat white plain they overlook. Slipping between the peaks and the surrounding islands, long white fingers of sea ice reached inland.

When people think of Greenland, they do not think of epic mountain ranges, mile-deep fjords, and icebergs as big as Madison Square Garden. They think of a territory on a 1980s board game or a splotch on a world map. Like Antarctica, Siberia, Arctic Canada, Alaska, and the last unexploited wildernesses on earth, the place is too big, too remote to conjure. If you look down at it from the window of a twin-propeller plane, the opposite becomes obvious. In the world of snow and ice, Greenland is everything.

The captain of our flight was kind enough to ease us into life in the Arctic Circle by setting the temperature in the passenger cabin close to the freezing point. I could see my breath as the plane lifted off from the fourteenth-century Icelandic port town of Hafnarfjörður, alongside the lava fields of the Reykjanes Peninsula and iced-over Faxaflói Bay. Heavy swells rolling through the deep blue North Atlantic looked intimidating even from thousands of feet in the air. It had been a stressful day of flying, for a few reasons. The rapidly spreading coronavirus pandemic was following me around the world. It had spread from China to the Italian Alps during my trip in the Dolomites and began to surface in New York City a few days after I got home. I wore a surgical mask for the first time on the flight from New York to Iceland. The plane was empty; I shared the economy cabin with four other passengers.

Two Americans sat a few seats ahead of me on the connecting flight to Kulusuk. The man was in his fifties and wore baggy cargo pants and a long-sleeved T-shirt. The woman was younger and had on a down puffer jacket and jeans. He was Scott, a fourth-generation New Hampshire skier and entrepreneur. She was Wendy, a massage therapist and radiologist raised on an Indian reservation. They were engaged, both launching blissfully into a second marriage and aggressively ticking off their bucket lists.

At home, Scott and Wendy spent fall days clearing ski runs at a backcountry ski area, where the National Forest Service allowed guests to thin trees from gladed slopes and cut skin trails for climbing. They like to ski there at night under a full moon (no headlamps). Their smart but thrifty assemblage of down, fleece, wool, and duct-taped Gore-Tex reminded me so vividly of New England winters that for a moment I thought that I knew them. I would get to know them well over the next week, as the three of us were headed out on the same Arctic adventure: a week-long dogsled expedition up the east coast of Greenland. The plan was to ride north with a group of Inuit seal hunters from Kulusuk for five days, running the dogs, sleeping in tents and hunting shacks, and assimilating with the oldest Arctic civilization on earth. With any luck, if the sea ice was thick enough, we would make it beyond the typical sledding circuit and breach *tunu*—a word that refers to remote parts of East Greenland as well as any frozen sweep of ocean, unclimbed peak, or "land out back" that few people ever see.

The trip was arranged by Pirhuk, a guiding and logistics company based in Kulusuk. Matt Spenceley founded the company after he began climbing and skiing in Greenland more than twenty years ago. A professional climber and skier, Spenceley had competed in the Ice Climbing World Championships and in 2008 logged one of the fastest crossings of Greenland: seventeen days and twenty hours to cover 450 miles between Nagtivit and Eqip Semia. He graduated from college when he was nineteen, served in the Royal Marines Commandos in Scotland, and now lived a life of perpetual ascent and cascade in Greenland.

Spenceley was a teenager when he first visited Kulusuk. The Utuaq family took him under its wing; Georg Utuaq, the father, showed him around the many mountains and fjords surrounding Kulusuk, teaching him how to travel across sea ice and open water and avoid polar bears. Spenceley learned to steer a dogsled, net seals, and survive on dehydrated food rations and Arctic char on backcountry expeditions that lasted up to a month. He befriended Utuaq's two sons, Justus and Mugu, and in keeping with Inuit tradition, was eventually taken in by Georg as a foster son.

Greenland had long attracted inquisitive, wandering mainland souls. The Greek philosopher Pytheas was the first European to see the island, or at least get close to it. He sailed north from France in the fourth century and wrote eloquently about pack ice and coastlines near Scoresby Sound in the Greenland Sea. An Irish monk named Saint Brendan arrived two hundred years later in an oxhide boat—in search of locations for new monasteries. Three hundred years after that, a Norseman named Erik Thorvaldsson (Erik the Red) was exiled by the Vikings from Iceland—for starting a landslide above his neighbor's farm, then killing the neighbor after he attacked Erik's slaves. To everyone's surprise, Erik did not sail east for Norway but headed west instead.

Ancient Icelandic sagas recount his endeavor at colonizing southwest Greenland. In 985, twenty-five longships followed him. Eleven were lost at sea in a storm; fourteen made it to shore. At its peak, five thousand Norse inhabited three towns on Greenland's east coast. Erik was chieftain of them all and prospered in his newfound home. His son Leif Erikson was ten when he was exiled with his father. Years later, the son would continue the family tradition of sailing against the grain by discovering the coast of North America, in what we know today as Newfoundland. (Erik the Red was supposed to accompany his son on the journey but fell off his horse on the way to the boat and took the sign as a bad omen. He died the following winter in an epidemic that wiped out most of the colonists.)

Norse settlers remained on Greenland for centuries, unaware that they were experiencing a mild climate event known as the Medieval Warm Period—which lasted from AD 900 to 1300. When the Little Ice Age descended at the turn of the fourteenth century, cropland was covered with several feet of snow and the colony starved to death—a morbid theme that would be repeated in various iterations as foreigners tried to settle the island.

"Don't Let It Ruin Your Vacation!"

Spenceley met us in the lobby of the Kulusuk Airport. *Lobby* and *airport* might not be the right words here. Air travel, along with many other things in Greenland, is a casual affair. The plane lands, the gangway drops, and passengers wander around the tarmac waiting for someone to open the cargo door. You can grab your bag there. There is also a chance that someone will load it onto a trolley and wheel it to a snowbank that doubles as the baggage claim. You can wait in the departure lounge, as I did among piles of seal and Arctic fox skins for sale, or the waiting area where a lovely but stressed-out older woman turned out prodigious amounts of french fries from a slick stainless steel deep-fat fryer.

I eventually found Spenceley talking with Scott and Wendy, who had managed the arrival routine far better than I had. Spenceley's wide shoulders, slim waist, long legs, trimmed beard, and magazine-worthy smile set the tone for the week: he and his guides were in charge; without them, we would all likely die in a matter of hours. It had happened before, not at Pirhuk, but on the island. Mainlanders had been testing their mettle in Greenland for two centuries, only to be chomped by polar bears, devoured by crevasses, swallowed by sea ice, starved to death, or killed by the all too common death-by-freezing.

The dangers here are so many, so diverse, and so extreme that a member of the Pirhuk team was required basically anytime you stepped outdoors. Spenceley's partner and lead dogsled guide, Rich Manterfield, explained this to us outside the terminal as Spenceley loaded our bags onto an ATV fitted with snow treads and a cargo sled. Manterfield slung a .30-06 rifle, encased in a blaze-orange sheath, over his shoulder and told us that we would be stretching our legs on a mile-long walk to the village.

The sky was white with a high, liminal haze, and the air was stunningly cold, even through a down jacket and several underlayers. The climate was not like at home, where you feel your body getting colder one degree at a time. You are just cold, immediately. We walked in single file behind Manterfield. Our job was to watch for polar bears, he said, on the tarmac, on the hills, on the path in front of us. It is no small irony that the cutest, fluffiest bears on earth also happen to be the planet's largest land carnivores. Weighing up to a thousand pounds, polar bears can knock a human head off with a quick swipe of their paw. They can also outrun you. They require twelve thousand calories a day. With sparse vegetation in their habitat, all their calories come from meat. If you see one, it sees food. Recent studies suggest that as the Arctic thaws, the bears' food web is being disrupted, ultimately increasing human–bear encounters.

"Don't let it ruin your vacation!" Manterfield said cheerily. He was the opposite of Spenceley in many ways. Skinny, boyish, with scruff instead of a beard. In a sled-dog pack, Manterfield would not be the lead dog. He'd be the scrappy husky in the middle carrying all the weight. He was thirty-four years old and had never dogsledded before coming to Kulusuk. He had been a climbing guide in the Dolomites, near one of the villages I skied through. He'd been university friends with Spenceley's wife, Helen, and joined Pirhuk in 2016. It became quickly apparent that Manterfield possessed a deep kindness and concern for pretty much everyone around him; he was studying to be a UAIGM (Union Internationale des Associations de Guides de Montagnes) guide, the highest certification for a mountain guide.

Manterfield practiced his guiding etiquette throughout our trip, brewing coffee before we woke up, warming our boots while we slept, eating last, cleaning up after us. He had been exploring the island with Spenceley and Helen for five years, learning from Inuit hunters and navigating trails, waterways, and dogsled routes. He reached a bit farther every trip, skiing remote mountains that had never been skied, trekking with groups in the summer to waterfalls and wildflower pastures. This was Pirhuk's business, and it required a tight balance of introducing guests to the area and Inuit culture without infecting them with Western money, influence, and, in the case of our arrival, disease.

I had scheduled a homestay with the Utuaq family for a few days before the dogsled trip began, but Georg had kindly asked that I refrain. The coronavirus was attacking older, compromised people. Georg was fifty-eight and had recently survived a six-month bout with cancer. With only two nurses on duty and a basic medical clinic, the village was susceptible to a major outbreak. Precaution soon became a theme of the trip, as villagers kept their distance and covered their mouths when we were around.

We walked past a graveyard marked with white wooden crosses, a few wreaths, and two shovels frozen in the ground. A cell phone tower loomed on top of a hill behind it. Everything in between was ice: ice slicks, frazil ice, grease ice, pancake ice. In the middle of Torsuut Tunoq Sound, dark blue icebergs were frozen in place. Where the ice reached from sea to land, giant cracks had opened up. Due west, smooth sea ice ran to jagged pack ice wrapping around the southwestern edge of the island.

The mountains cradling the Apusiaajik Glacier obscured the horizon to the north with a line of peaks that looked like they could have been plucked from the Alps. Behind them, several subranges cut a jagged line across the sky on the far side of the Ammassalik Fjord. Manterfield told us to move to the side of the trail as a villager approached on a dogsled. The sled ran on thin, curved wooden runners, a lightweight frame lashed together with nylon, two curved handles, and a steel brake in the back. Manterfield said that locals used dental floss to stitch nylon strapping and that the sleds were built to flex as they slid over bumpy terrain. Different parts of the island used different sled designs. In the mountains, the sleds are shorter, to maneuver tight turns. In the north, they are longer to cross sea ice without breaking through. Sled makers import entire trees from Denmark—there are only a handful of forests on Greenland—to use for runners and a frame. The frame has to be perfectly symmetrical, or it will break up on the incredibly rough and icy ground.

Greenlanders steer from the front of their sleds after they get the dogs going, unlike Alaskan Inuit, who steer from the back. The middle-aged man driving the sled past us sat sideways, one leg sticking out over the snow, the other casually crossed. You could barely hear the dogs or sled as it slid by. Still some distance from town, we passed a rusty diesel tank the size of a two-bedroom home. Manterfield said that hunters climbed it every morning with binoculars to look for polar bears. On the right was a maintenance garage and three bright red shipping containers stacked with food and supplies. On the left, the town store, the epicenter of Kulusuk, where you could get Danish pastries, antifreeze, blue jeans, and milk. (Most residents paid with a monthly welfare check sent by the Danish government.)

The store had it all, Spenceley said later that night. Maybe too much. Some of the old ways of living, subsisting, traveling across ice, and surviving winter were slipping away. Snowmobiles threatened to replace dogsleds. Cell towers delivered social media, Amazon, and all of the comforts of the Temperate Zone. Locals walked through town wearing synthetic down jackets with cell phones pressed to their ears. Kids played video games in their rooms and surfed the internet to see what the rest of the world was doing.

Hunting had been relegated to a hobby, Spenceley said, as the store provided most things villagers needed. The last hunters were proud of and good at what they did. They took on apprentices as young as seven or eight years old. It was a way to save the culture, to keep the next generation speaking the language and learning traditions. They chipped breathing holes into the frozen fjord and set nets to snare seals. They also hunted seals with .22 rifles in patches of open water that wobbling icebergs pound through sea ice. Once it gets cold enough and ice has spread through the inlets, villagers can zip over it along the coast on dogsleds, hunting seals and bear and sleeping in tiny hunting cabins.

Walking into Kulusuk... A basketball court surrounded by a chain-link fence sat in the center of town. Candy wrappers and empty cans of Carlsberg beer littered the ground. Two kids played on a small, wooden playset at the top of the hill behind the school. They had hauled a bright yellow sled to the top of the slide. One kid climbed in; the other watched. A pack of dogs nearby howled and walked in circles. Helen said that teachers were starting to tell the children to pick up litter around town. They know about climate change, she said, they just think they can't do anything about it. They might be right about that.

The Inuit do indeed have more than fifty names for snow, though the words refer specifically to different types: fine snow, deep snow, snow on the ground. They have just as many terms for ice: sea ice, freshwater ice, ice on the inside of a tent, pressure ridges, icebergs in the water, icebergs frozen in the ocean, melting ice, ice floes. The words were derived by necessity, as their ancestors learned to live on the ice. The difference now is not so much that the Inuit have radically changed; it's that the world around them has. Invisible atmospheric rivers had changed course; deep ocean currents delivered warmer water to the island. It wasn't just Greenland. *Arctic amplification* was warming the top of the world faster than anywhere else on the planet.

Amplification is largely due to changes in albedo, as dark earth and water are revealed when snow and ice melt. It has warmed northern regions like Alaska and Greenland at twice the rate of midlatitude areas. Warmer water delivered by the Gulf Stream and the North Atlantic Drift exacerbates warming, as do myriad other phenomena, such as an increase in thunderstorms at the equator; the storms cycle hot air into the upper atmosphere and eventually to the poles. A 2020 study from the University of Copenhagen concluded that Arctic amplification has been vastly underestimated; researchers noted that temperatures over the Arctic Ocean have risen by 1 degree Celsius every decade since 1980. Over the Barents Sea and around Norway's Svalbard archipelago, the increase has been 1.5 degrees Celsius per decade over the same period. (Even Antarctica, whose landmass is always colder than its northern twin, is warming at three times the global average.) The conclusion from these observations? The Arctic has not been as warm as it is now in millions of years.

The first 100-degree Fahrenheit day in the Arctic Circle was recently recorded in the Siberian town of Verkhoyansk. Longyearbyen, Svalbard, the world's northernmost inhabited settlement, saw multiple days of 60- and 70-degree weather in 2020. In the Canadian Arctic, temperatures in Eureka, Nunavut, located at eighty degrees north latitude, hit 71 degrees Fahrenheit the same summer. The effects on Arctic forests are similar to those in Kim Maltais's backyard in the North Cascades: warmer temperatures dry out the soil, resulting in record wildfires. Eastern Siberia lost fifty million acres of forests—about the size of Greece—to fires in 2020, emitting nearly two hundred megatons of carbon dioxide into the atmosphere and thawing methane-rich permafrost and peat soil. The result has been what climatologists call *abrupt warming*—increases of more than 3 degrees Fahrenheit per decade in some Arctic areas for the past forty years.

The people of Kulusuk have seen similar changes. Winter there now starts several weeks later and ends sooner. One of the fastest-moving glaciers in the world, the Helheim Glacier, which sits at the head of Sermiligaaq Fjord, now flows at more than seventy feet a day, dumping millions of tons of ice into the ocean. (Visitors can watch it slide by underfoot.) To the north, the Kangerlussuaq Glacier—which accounts for around 5 percent of all ice discharged from the Greenland Ice Sheet—retreated three miles in just two years, between 2016 and 2018. The island lost the equivalent of an Olympic swimming pool of ice every second in the mid-1990s, or around 25 billion tons of ice. In 2020, scientists found that number had increased ninefold, to 234 billion tons per year.* Michaela King, who flew to JIRP's Camp 18 with me, was the lead author of a 2020 study that showed how the ice sheet had passed a point of no return. Glaciers were sliding into the ocean faster than snowfall could replenish them—which would mean another of the planet's nine tipping points had fallen.

Sea ice isn't faring much better. In July 2018, the Arctic lost more than forty thousand square miles of sea ice, an area larger than the state of Kentucky—*every day*.† Researchers projected that summer sea ice in the Arctic Ocean will be completely gone by 2030 for the first time in two million years. Shore-fast ice—which forms along

* Greenlandic ice melt from 2019 alone would cover California in four feet of water.

† Similar to snow cover, sea ice, as it melts, reveals dark, open water that absorbs more than 60 percent more sunlight than does ice-covered ocean. In the month of September alone, sea ice cover has dropped more than 12 percent per decade between 1979 and 2018.

the coast in shallow water—makes up around 12 percent of sea ice cover and is essential to Arctic communities for hunting, communication, travel, trade, and survival. A 2020 study found that spring melt of shore-fast ice will start up to a month earlier by the end of the century, melting out ancient transportation routes—including the trail we were now supposed to follow.

When Spenceley first arrived in Greenland in 2000, Ammassalik Fjord still occasionally froze halfway up the inlet. (Georg Utuaq grew up hunting and fishing on the fjord.) Now it was open water all year long. Inlets along the east coast were melting out too, stranding communities and forever altering a way of life. The route we were planning to dogsled had been open water two weeks ago, Manterfield said. But a sudden cold snap had finally frozen it. "It should be thick enough for us," he said, looking out at a large pool of seawater in the middle of the bay. "I guess we'll see, right?"

15

A Brief History of Death and Survival

With the homestay off, I was confined for two days in the brand-new, remarkably comfortable Pirhuk lodge. Spenceley, Helen, and Manterfield built the structure over the previous three years. Tales of the construction sounded like war stories: chaos, confusion, fear, sadness, disappointment, and, ultimately, extreme relief and joy. The thing about living on a remote island is that pretty much everything—food, lumber, Sheetrock screws—has to come by ship. The Royal Arctic Line supply ship carrying materials to finish the lodge was crushed and disabled by pack ice, forcing the Pirhuk crew to brave the seas and transport their stuff with a small launch until another boat from the Faroe Islands arrived to offload the cargo. The local carpenter they hired to work through the winter—while they returned to the mainland—assembled much of the plumbing, the doorways, the electrical circuits, and a hand-hewn table that Helen had spent weeks planing and sealing—backward. Storms soaked and tore apart any structure that was not covered. Cold and dwindling light in the fall made work conditions miserable. The crew members had to simultaneously do their guide work to pay for the new lodge while everyone—crew and guests alike—was accommodated in a tiny house in town.

A combination of caffeine, forced smiles, wine, and an indefatigable passion for Greenland and adventure pulled them through.

The result was a tidy refuge with central heating, dual bathrooms, hot showers, a shiny commercial kitchen, private rooms, and a handsome dining hall provisioned with coffee, muesli, and fresh baked bread every morning. The cook was a soft-spoken artist named Theresa, who sailed to Greenland once on a tourist ship and never left. (One-way trips like this were common.) Along with Helen, Theresa kept the place spotless and cooked tremendous meals every night, accompanied by box wine and cans of beer in the pay-as-you-go Honesty Bar.

I was enthusiastically honest my first night and was happy to settle into the sheepskin-lined lounge the next day. Wendy and Scott had booked two backcountry ski touring days with Spenceley and shuttled frantically back and forth from their quarters to the gear room—jet-lagged, sniffling, trying to assemble avalanche transceivers, climbing skins, gloves, goggles. Through a picture window, Iperajik Mountain seemed to lift off the sea ice like a spaceship, topping out at twenty-six hundred feet. It was twenty degrees Fahrenheit outside and seventy inside, thanks to a diesel furnace humming in the corner. Six hours later, I looked up from a pile of books about Greenland's first settlers with what I can only describe as a completely new understanding of winter.

An important distinction between winter in Greenland and winter that most of us have experienced is that an Arctic winter never really ends. Or it does, for a month or two, but freezing temperatures are never far off. The Inuit word for *winter* also means "one year." When the Northern Hemisphere tips away from the sun, time becomes abstract. Nights last for eighteen hours. Dawn and dusk merge, as the sun never fully rises above the horizon. Most Greenlanders don't recognize the difference between day and night. They tend their dogs at midnight and have friends over for coffee at three in the morning. "My foster brother called me at one in the morning last winter," Spenceley said, "to see if I wanted to change the oil in his outboard engine with him. I have to turn my phone off at night."

It never really gets dark in the Arctic, either. The reflective surface of the snow and glaciers glows blue under the moon and stars. Night is an afterworld of ghostly shapes and distant sounds, like opening your eyes underwater. Inuit legend says that this is how the world began, at night with no light at all. No one died in that realm. When people multiplied and earth became overcrowded, light came again and humans began to die.

Winter in the Arctic is a time to contemplate—how the earth was made, how we were made, the balance between human existence, place, and climate. Weather is more than a conveyor of wind and snow to Inuits. It is a living being with a name and a personality. Selah controls both weather and consciousness. Whatever the weather is doing, you are doing: walking north, sledding south, hunting musk ox, netting seals. Weather is fate, and as a consequence, the weather gods are depicted as ruthless, sometimes even treacherous.

Starvation was common in Inuit villages in the old days. Senicide (the killing or abandoning of elderly people) and infanticide were practiced on extremely rare occasions, such as during a famine, to keep a village alive. If the sea ice didn't come in, whales couldn't feed off bait living beneath it and never showed up. If musk ox or caribou deviated from their migratory routes, a village would be left without food for months. Infants were suffocated to end their suffering. Family members "mercy killed" grandparents with a dagger or by pushing the old souls out to sea on an ice floe. This practice was not tragic; it was pragmatic. It was survival, fate. Nature is the dominant force in the Arctic; humans live at its mercy. "People here think it is silly to be afraid of dying," Spenceley told me one night at dinner. "My Greenland family would say, 'What did you think was going to happen?' "

Living so close to winter and death formed detailed and sophisticated Inuit mythology that exists in multiple, overlapping realms—not distinct dimensions like heaven and earth. The borders of these realms are transition zones, where ice turns to ocean, flatlands rise into mountains, humans morph into animals, and animals become spirits. Inuit shamans say that sun dogs—rings around the sun—are the head of a drum that can be beaten to invoke another world. Stars are windows in houses across the land of the dead. The moon absorbs spirits of the dead and transports them there. Another myth says that the stars are holes in the sky from which snow, rain, and spirits fall to earth.

Serratit is the old world, where sorcery could bring the ice back and return things to the way they used to be. Serratit is fantastic, dark, comedic at times. The moon and sun were brother and sister then. They slept together before they knew they were related, and when they realized what they had done, they launched into the atmosphere to live in different houses—the sun leaving her house in the summer, the moon emerging in the winter. Selah, the weather god, is a baby giant who fell to earth from the hood of his mother's anorak. Women gathered around the baby and played with its penis. It got so large it took four of them to hold it down. Then the baby rose into the clouds, removed its caribou skin diapers, and let loose with wind, snow, rain, and the seasons. To control the weather, a shaman had to soar into the sky and refasten the diaper.*

The lives of animals and humans are deeply intertwined in Inuit mythology. In past worlds, humans and animals could speak to one another. They could also change bodies and make love. Bears absconded with women; men married dogs. For a shaman to recite

* In varying interpretations of Selah, or Silla, the deity represents the life force of "everything," similar to the Polynesian concept of mana, or what Emerson called the "Over-Soul": "a power / That works its will on age and hour."

these old beliefs, it had to be early in the morning and the hood of the shaman's jacket had to be pulled up. The speaker would then jam the ring finger into their throat and gag, after which the old words emerged. A physical manifestation of the magic is a *tupilait,* a small figurine made of bones, hair, and seal skin, along with clumps of moss tied with sinew. The figurines are placed in a sacred spot. Then the speaker puts on a jacket backward, with the hood pulled up over their face, recites a song, and rubs the bones against their own genitals, thus giving the *tupilait* life.

If there is no ice, it is because the sea goddess, Nerrivik, had lice in her tangled hair and no one had combed it. Nerrivik had one of the closest relationships with humans. She was orphaned as a child and left behind by hunters to fend for herself. When she swam to their kayaks and tried to pull herself up, they cut off her fingers. The fingers became seals, walruses, whales, and other marine life that the Inuit depended on and that she became guardian of. If humans disrespected the animals the way they had disrespected Nerrivik, she hid them under her skirts and let the humans starve. She could only be appeased by a shaman traveling to her house at the bottom of the sea and stroking her tangled hair.

Most of us think of Inuits and other indigenous cold-weather groups as people who live directly north of us. Few think about them as a contiguous culture that wraps six thousand miles around the top of the world. Their eastward migration from Siberia, where they originated, came in pulses. The first pulse was from Chukotka, Russia, a football-shaped chunk of land on the Bering Strait.* Early Saqqaq voyagers crossed the strait six thousand years ago. They weren't looking for farmland or river valleys; they had broken away from temperate regions and populations to embrace

* Remarkably, the eastern half of our civilized world is divided from the west by just fifty-one miles, a few days' walk or paddle for a proto-Inuit family.

winter and unsettled lands in the north. As civilizations to the south experimented with large-scale agriculture, urban development, and the rule of law, bands of Saqqaq turned east to find reindeer, caribou, musk ox, and, later, seals, walruses, narwhals, and polar bears.

Clusters of families eventually breached Alaska's Brooks Range and crossed from Arctic Canada to Ellesmere Island, Smith Sound, and, finally, Greenland, arriving around 2500 BC. From there, some followed herds of musk ox to the northern reaches of the island; others headed for the fjords and open water on the south coast to live off the ocean. Recent genetic evidence shows that the Dorset culture, which spread across the Arctic following the Saqqaq, were of the same lineage, though from a different migration with a distinct lifestyle. The Dorset culture depended largely on hunting marine mammals, typically finding them in breathing holes in the ice. They were master carvers and left behind three-sided harpoon heads and intricate amulets and masks.

Direct ancestors of today's Inuit, the Thule people, arrived in Greenland around the first millennium. They hailed from the coast of Alaska and were an advanced and versatile culture, using kayaks to hunt and fish, knives shaped from slate, and harpoons with buoys—inflated seal stomachs and intestines—to take down large prey like whales, walruses, and narwhals. The Thule people might have had contact with, and even traded with, Vikings descended from Leif Erikson and other explorers of the eleventh century.* They probably coexisted with the Dorset people, who the Vikings described as larger, gentle, and shy. Over the next several hundred years, the Dorsets died out as the Medieval Warm Period spiked temperatures 1.8 degrees Fahrenheit, melted sea ice, and made it impossible for them to hunt.

These changing temperature patterns in Greenland and their

* Vikings refer to the Thule as Skrælingjar in epic Norwegian poems.

effect on various early cultures' survival are other examples of how volatile earth's climate is and how civilization is often too rigid to withstand this volatility. Unmovable things like glaciers, the ocean, rivers, and forests can shift in a matter of years, dumping a truckload of Jon Riedel's boulders on the proverbial avalanche field of earth's ecosystems. An interesting difference between the hunter-gatherers of yore and the organized, sedentary societies of today is this: hunter-gatherers were more versatile, and they adapted to climate change quite skillfully.

Home Rule

It was noon when I stood up, stretched my legs, and stepped outside for a walk around town. At least, I thought it was noon. I really had no idea, since daytime in early March was already fourteen hours long. The sky outside was the same color as the snow, milky white, creating a singular plane through and above the village. The mountains were faint imprints on the horizon, three-billion-year-old gneiss formed by lava flows during the Pliocene. I developed vertigo and lost my balance five steps outside the building, looking down at the snow, then up at the sky.

Helen allowed me to go alone, as long as I promised to get indoors if I saw a bear. Hunters had spotted one nearby a few days ago; I walked past its outstretched hide two hundred feet into my tour. The mouth was pried open with a stick. Its baseball-mitt-sized paws were yarded down with rope to a latticework of wooden rafters.* Some of the houses in town had concrete foundations; others

* It is illegal to export polar bear pelts in Greenland; most end up in a Kulusuk home. The person who spots the bear typically gets the hide. Whoever shoots it gets the heart and first pick of the meat. The person who field-dresses the carcass and hauls it back to town gets the second pick (usually the rear haunches).

were supported by pressure-treated pylons and joists. You could tell which ones were not insulated by the curtains of icicles growing from the eaves. Shirts and underwear hanging from clotheslines were frozen solid. Upside-down dogsleds and yellowed ten-gallon water containers sat in a few front yards. Two galvanized exhaust pipes jutted from a shed near the store, where a roaring diesel generator provided electricity for the town.

The school building looked empty. A fiberglass dory sat high and dry next to it. Five children wearing one-piece snowsuits had carved a 300-yard sled run from the top of a ridge, around a dozen houses, and down to the store. The fluorescent green and yellow sleds were

This bear was brought in on our last day in Kulusuk. Inuit hunters consider bears the most intelligent of prey, almost human-like. After the bears are killed, the hunters offer male bears knives and female bears skin-scrapers and needle cases, then hang the furs in their homes to honor the beasts. If the bear is treated well, the hunter will be able to kill others. If not, bears will stay away.

advanced—with steering wheels, skis that turned, and racing stripes. A few dozen Greenland dogs howled as the kids flew down the track. A quarter of the houses kept a pack of dogs tethered in their front yard. The beasts walked in circles stained yellow and brown by their business. They whined constantly. They were fed once every three days so they didn't get used to a regular diet and could handle being out on the trail with slim rations.

Kalaallit qimmiat were brought to the island a thousand years ago with the Thule people. They are a bit smaller and lighter than Alaskan huskies. Like most native dog breeds in North America, huskies descended from Eurasian gray wolves, meaning Greenland dogs probably migrated with their masters from Siberia. Archaeological digs in Siberia have found 8,000-year-old canine corpses buried with weaponry, jewels, and their masters. A 2,000-year-old knife handle from Ust'-Polui depicts a dog hauling a sled.

Today, Greenland dogs are as important a cultural icon as are language and traditions like hunting and fishing. The dogs are microchipped and logged in a database when they are born. It is illegal to import dogs anywhere on the island except in the southwest. Still, along with many Greenlandic traditions, the number of dogs continues to dwindle, hitting an all-time low of fifteen thousand in 2016.

For wild animals only a few evolutionary steps away from a wolf, the dogs are surprisingly docile. They arch their backs and hang their tail low when you approach, then perk their ears and dip their noses. When they lie down, all four legs stick straight out, like a cartoon dog slipping on a frozen pond. They don't sleep inside or on a pad. They curl into a furry ball on the snow, tuck their muzzles under a hind leg, and nod off in temperatures well below zero.

I wandered between dog packs, looking for the western edge of town and perhaps a glimpse of Ammassalik Fjord. Two local boys passed me with their baby sister in a wooden sleigh. The boys wore down jackets with synthetic fur hoods. Two others straddled a banister outside the school, giggling and whispering to each other. Above them, on a small hillside, was another simple graveyard with a few white crosses.

A red Danish government helicopter made endless laps between the airport and surrounding villages, delivering supplies and transporting people. Greenland has existed as a colony of the Danish Kingdom since the 1600s, when King Christian IV sent ships and soldiers here. Greenland officially became a colony in 1721, but since 1979, it has enjoyed home rule, which grants it significant administrative control. In 2008, Greenlanders voted for self-governance, giving the nation total control, over time, of virtually everything except military and foreign affairs.

I crested a small hill and saw the snowscape vanish into the sea. Here was the ice that you see in magazines, on book covers, in Grey's favorite cartoon—in which woolly mammoths and prehistoric squirrels save the planet. (They are vastly more effective than humankind.) There were ice sheets, icebergs the size of cruise ships, smooth shore-fast ice, sea ice wrapped around a circle of deep green ocean, a sea of shattered pack ice moving in from the southeast...

"Glacier ice," a man in a neighboring yard called out. "The current brings it here." He swept snow off his dogsled, preparing it for a regional race to be held in Kulusuk the next day. He announced himself as Lars Bianco, right hand on his chest, left hand on a five-foot wood and steel ice pick used to test sea ice thickness. I would later learn that Lars was a colorful figure in town—an expert hunter who was a little less expert, or unlucky, at the helm of a dogsled. At the regional competition the year before, he was still harnessing his dogs at home when the starting flag dropped. His dogs heard the start and raced through town at top speed, with Lars barely hanging on. Then they ran the full length of the starting line sideways twice before launching into a jumble of pack ice—sending

Lars down a twenty-foot crack between two slabs. Spenceley and Manterfield went after him on a snowmobile, but he had already climbed out of the hole. It took three days to find his dogs.

I asked Lars about polar bears—still very much at the forefront of my mind—and he told me that they were out on the pack ice, hunting seal on the edge of the open ocean. The ice was too thin for Lars to hunt, he said. He was waiting for it to melt so that he could launch his fishing boat. "September, October, November," he counted, unfolding one meaty finger at a time. That was the season he could count on these days: open water and lots of fish. I thanked him for the chat, and he held up a hand and said *quyana,* "thank you" in Inuit. I thanked him again, and he held up his hand and repeated *quyana*. Then he pointed to me. *Quyana,* I said, and he grinned and went back to sweeping his sled.

A raven swooped from streetlight to streetlight as I walked back to town. A few patches of blue sky opened up. Clouds touched the tops of the surrounding peaks. Where the wind had blown the snow away, brown tufts of alpine fescue and dwarf harebells fluttered in the wind. Manterfield and Helen drove past on the ATV to pick up more guests at the airport. Seven ravens perched on top of a shipping container watched me walk by. Below them, two speedboats sat in the backyard of a house, waiting for the ice to melt.

"There Is Only Maybe"

That night, Manterfield gave us a briefing. A major weather system was spinning off the North Atlantic and was headed toward the east coast of Greenland. He was going to postpone the trip for a day to wait it out. The storm was a classic North Atlantic cyclone, the kind that had sunk hundreds of ships off Greenland's shores. It was two hundred miles wide and would bring northeasterly hurricane-force winds. Since we would be heading northeast for the first two days of the trip, the danger of frostbite and hypothermia was significant.

It was a fairly common scenario in Kulusuk, Spenceley said. The island's icy bulk and its location near the top of the world acted as a blockade to powerful westerlies, which have to go around Greenland's rugged coast instead of over it. Like the Bernoulli effect on an airplane wing, these so-called tip-jet winds accelerate up to ninety miles an hour as they bend around the island. The wind's cooling of the water off the coast can shift major ocean currents—ultimately changing how heat and moisture are distributed across Europe, Asia, and even the west coast of America. A few weeks before we arrived in Kulusuk, an extreme low-pressure bomb cyclone developed east of the island. It was so strong it sucked an entire low-pressure system that had been drenching the US Southeast coast for days and slung it across the Atlantic over Europe, causing flooding on European coasts, deep snow in the Alps, and a few dozen commercial flights from the US to break the sound barrier for the first time ever, riding a 250-mile-per-hour tailwind.

The room was awkwardly silent after Manterfield's speech. The wind was already ripping outside, rattling windows. Swirls of snow whipped down the main road in town. I wondered if everyone on the expedition knew what they had gotten into. Scott and Wendy seemed hardy enough to survive anything. The new guests that Helen and Manterfield had picked up, Jason and Mayah, were a powerfully optimistic San Francisco couple who seemed to be grappling with their initial excitement about sledding for a couple hundred miles with what Mayah called "doggies." In an action thriller about a pandemic wiping out life on earth, Scott and Wendy's role would be on the front lines, leading a fearless charge to save their town, then dying heroically in the process. Jason and Mayah would have been sacrificial lambs from the opening scene, until they shocked the audience with their unbelievable resilience and smarts—

allowing them to survive and *grow together,* simultaneously conquering their own fears, refilling the common well of love they shared for each other, and emerging as the sole survivors of planet Earth who would then seed a new, equitable, gorgeous, and socially progressive civilization.

I got to know Jason and Mayah a bit more the next morning on a snowshoe hike with Manterfield. Pirhuk sells its trips as "expeditions," necessitating that we get up at dawn, force down food and coffee, pack an unnecessary amount of water, and walk into the Arctic at seven in the morning on what would amount to an afternoon stroll. Fifteen minutes into the hike, Mayah appeared to be freezing to death, and Jason did not seem comfortable with the idea of snow. I had never snowshoed before and was delighted to find that the contraptions not only floated in deep snow but also provided incredible lateral support for my knee. I led the way up a meandering snowmobile track. The objective was a small, barren peak about two miles away. A hazy sun hung above the mountains. Birch saplings four inches tall grew around the rocks and windblown ridgelines. There were no full-sized trees. Small heaps of igneous rock broke through the snow every hundred yards. The pitch steepened the farther we walked until it leveled out on a low ridge a mile from the lodge.

Manterfield told us that a US Distant Early Warning radar base operated there in the 1950s. It was part of the same military effort that brought JIRP's Malcolm "M3" Miller to Greenland in the opening days of the Cold War. M3 had been commissioned to find a way to launch missiles through sea ice. The Kulusuk base was built to watch for ones headed in the other direction. Soldiers were not allowed to mingle with Inuit villagers, Manterfield said. Yet over a decade, mixed-race children appeared and some locals began speaking English. You can still see the influence of the base at Kulusuk's dance hall today, Manterfield said, where swing dancing is the main attraction. "The boys stand against the wall and the girls walk around in a circle, deciding who they want to dance with," he said. "If the DJ plays rap or techno, the floor is empty. When he puts on a swing tune, even the old folks who have a hard time walking get up and dance."

Forty minutes later, we crested the summit. Wind scoured the peak and froze bare patches of skin on my cheeks, as the first swirls of moisture from the coming storm obscured the horizon. Every phase of frozen water was on display below: snow, shore-fast ice, sea ice, open water, pack ice. Across Ammassalik Fjord, the Helheim Glacier continued its terrible slide. The wind was too loud to speak over. We ate a quick snack at the top, hoods up, backs to the wind, shoving food down so we could get going again.

The descent was fast and warm in the lee of the mountain. Mayah and Jason told us about the wedding they were planning.

The view of Ammassalik Fjord from the summit. The first written description of Greenland was in 1075, when Adam of Bremen included it in his Historia Hammaburgensis Ecclesiae. *It was first documented on a map in 1424, but it was the epic hunt for the Northwest Passage—which consumed captains and merchants*

They had rented a riad in Marrakech. Welcome cocktails would be served at a palace inside the city. The rehearsal dinner would take place at the Churchill Bar, where the old man occasionally reposed in the 1950s. I told them about a riad I had stayed in years ago. I was on a solo quest to hike and ski the tallest peak in North Africa, sixty miles east of Marrakech in the High Atlas Mountains. On a ninety-degree day in the dusty souks of the city, the icy summits looked like a mirage floating in the sky. For a hundred dollars, a taxi driver took me to the mountains in a 1972 Mercedes Benz. All I had brought was an old canvas backpack, rain pants, and a down jacket. A guide I met at the foot of the mountains lent me boots and a pair of old touring skis. The boots were two sizes too large but fit well enough. We hiked four thousand vertical feet through terraced fields and walnut orchards, then sat on milk crates and pillows with

for about three centuries after the discovery of America—that really put Greenland on the map, initially by British captain John Davis. First contact in the north soon followed, wiping out many tribes with disease.

a shepherd and ate hot bowls of tagine that he had cooked over a fire. After the snack, we skinned another four miles to a French climbing refuge built in the 1930s, and after a short, very cold sleep, we woke at dawn and summited 13,654-foot Jebel Toubkal. The Sahara consumed the horizon—a dark brown streak of rock and sand broken by a few walled Berber settlements and a half dozen green oases hugging a skinny river. To the west were the palmeries of Marrakech and one of the most endangered watersheds on the planet, as snow in the High Atlas melted at an astonishing rate.

Jason and Mayah got into a groove on the way back, and we made it to Kulusuk in time to watch the dogsled race. A quarter-mile-long starting line had been stamped into the snow. Six trails led away from the line and merged into one track a mile out. A man standing on an iceberg held a red flag over his head, then a yellow one, then green. My man, Lars Bianco, had made it to the line and looked ready to go. The favorite and two-time East Greenland champ, Justus Utuaq, was tangled in a mess of howling dogs. When the flag finally dropped, a young musher named Mikael Kunak got out in front first.

The whole procession ran hard for a mile, rounded a corner into a field of icebergs, then disappeared from view. When they returned an hour and a half later, Mikael had kept his lead and Justus had climbed back into third place. When Mikael crossed the finish line fifteen minutes later, his friends and family surrounded him and lifted his sled into the air—before the dogs hauled it and him away again. Bianco finished fourth, just qualifying for the next round and knocking fifth-place finisher Mugu Utuaq into last place.

Mikael had eaten dinner with us the night before. He was twenty-eight and was extremely mild-mannered. He rolled his shoulders when he smiled and went to great lengths to avoid eye contact. Because his English was limited (and my Greenlandic—other than *quyana*—nonexistent), it was difficult to strike up a conversation. A friend who came with him was more outgoing. Without a word, the boy picked up an acoustic guitar from the couch and lit into "Folsom Prison Blues," by Johnny Cash. There was no intro or explanation. He strummed a verse, then lit into a wild interpretation of the tune, with a blend of English and Greenlandic.

When I asked Mikael that night if he thought he would win, he casually said no, he would probably place third. Spenceley explained later that winning, success, and achievement were not things Greenlanders thought about much. "There is no 'Yes, I am going to do that,'" he said. "There is only maybe. We will see. If the conditions are right and things line up, perhaps it will work out." A life lived in pure winter, at the mercy of Selah, tempers pretension and creates a kind of nihilist-Buddhist philosophy among the Inuit. They don't understand why visitors constantly complain about the weather; it is what it is. They don't get why foreigners get worked up when they hit their thumb with a hammer. "Why would you be afraid of doing something dangerous when you knew it was dangerous when you started?" Spenceley asked. Which is to say that living in the frozen wilds of the Arctic Circle, Inuits know this world is brutal, perhaps more than any other population does, and so they know that human life is expendable, fleeting, insignificant even.

Back at the lodge after our short snowshoe trek and the dogsled race, Spenceley and Manterfield glanced outside at the weather again. It was not good; the storm had not weakened. That night, the windows sounded as though they were going to blow in. The sky was black with no stars or moon, and the corrugated steel roof hummed. I felt a cold draft blowing through a seam in the wall and bundled up in a comforter before falling into a deep sleep. In the morning, the full force of the cyclone had apparently arrived. The village was obscured by a whiteout. The wind roared and made the lodge timbers creak. I found Spenceley looking out the dining room window, and I lamented that we would have to delay the trip further.

"This?" he asked. "This is nothing; you're leaving in an hour."

16

The Lark's Foot

By the time Manterfield led us into the gale, Jason and Mayah had dressed and undressed four times. Scott and Wendy had been geared up for an hour, had sweat through their base layers, then taken half their clothes off again to dry out. The tiny gear room smelled like sweat, sunscreen, and the unfortunate fog of anxiety-induced flatulence. Shafts of cold, gray light spilled through a small window. The wind sounded like a NASCAR race outside the front door. Even Manterfield was jumpy. He checked everyone's zippers, goggle straps, food rations, and water bottles. Jason and Mayah's continuing downward spiral spooked him, and at the last minute, he demanded that everyone wear survival suits as a final barrier against the elements. The suits were waterproof, airtight, lined with fleece, and heavy as hell. Used primarily for a fishing boat crew whose ship was about to sink, the suits were so restrictive that the wearer had to trade the use of all four limbs in exchange for warmth.

Scott was red-faced about being ordered to put the survival suit on, even more so when I got a reprieve and was allowed to stash mine on the sled. The group walked straight-legged and straight-armed to the inlet. Three dozen howling Greenland dogs waited near the shore, along with three hunters and their sleds. I would ride with Justus and Manterfield. The champ, Mikael, would drive

Mayah and Jason, and Scott and Wendy were stuck with the moody Mugu, still brooding over his loss. (Spenceley had instructed us that under no circumstance were we to mention race results on our journey.)

Justus had me hold his sled while he untangled his dogs. My job required stepping on a steel brake the size of a bear trap to keep the dogs from yanking it away, perhaps forever. Because Justus was working in front of the sled, failure would also mean running him over. He wasn't worried. Justus was the oldest of the three hunters and had taken on the responsibility of getting us onto the ice and back home safely. He was the older-brother type—strong, watchful, slightly nerdy, serious. He had a round, chubby face and a small paunch and wore clear Oakley glasses when driving the dogs. His hands were like vise grips. He pulled apart knots so frozen and tight the rope seemed to have melded into itself. He lifted and threw around the sled, which was loaded with two hundred pounds of gear, as if it were a piece of driftwood. He always worked with bare hands—in a gale, underwater setting a seal net, at night when it was twenty below zero. (I caught him more than once glancing back to see if I noticed this show of fortitude.)

Greenlanders run their dogs in a fan formation instead of the neat pairs Alaskan mushers use. (The pairs allow Alaskan teams to navigate narrow forest trails; Greenland dogs can spread their weight out on thin sea ice in the fan.) The fan also creates a complicated arrangement of hormonal chemistry between males, females, and rival dogs; a driver can experiment with this chemistry to get the most pull out of each animal. The fan formation also comes with the complete chaos you would expect from twelve mostly wild dogs tied to one another—trace lines intertwining, paws tripping, dogs being dragged behind the sled. All twelve trace lines come together in a cinch knot called a lark's foot, which allows the driver to shorten all the leads, pull the tangled mess through, undo the knots, and then let them back out—all while running five to ten miles an hour over rock-hard ice. (Again, with no brakes.)

Each driver steered this slightly controlled state of entropy not from the back of the sled, where the brake is, but up front, setting up exciting and often terrifying driving-without-brakes situations. Manterfield told me to hang on tight at all times, as there would be no slowing down for an ice bulge or a snowdrift. Burlap sacks of dog food and other supplies hung from the handles on the back of the sled, along with our backpacks and a rifle. Justus and Manterfield sat on sleeping pads and reindeer hides as I reclined on three duffel bags behind them. Next to me was a forty-pound bag of raw seal meat for the dogs. Justus steered the sled by hollering "left" (sounds like "You! You! You!") and "right" (a high-pitched trill). He also hollered repeatedly at a dog named Apalartoq (Red), the same nickname the hunters bestowed on Manterfield years ago—referring to a second-degree sunburn the guide had suffered on his first climbing expedition in Greenland. This yelling of Manterfield's nickname over and over, all day long in an often derogatory way, was symbolic of the odd dynamic between the hunters and our guide. Namely, that Manterfield was in charge and the hunters didn't believe this to be the case.

About five seconds after Justus told Manterfield and me to jump onto the sled, the pack took off. We grappled for something to hang on to as Justus leaped onto the front of the sled. The dogs seemed to know where to go. We followed the race course for the first two hours, wandering north and east, past remnants of the Apusiaajik Glacier, now deep blue icebergs frozen in white sea ice. Taatsukajik Mountain rose to the right in crumbling ridgelines and thick, white blankets of snow. We rounded a corner and saw the disintegrating tongue of the Apusiaajik—a thin blanket of ice rising vertically from the head of Torsuut Tunoq Sound. Two hundred years ago, the glacier was the dominant feature of the inlet. Massive flanks encroached on neighboring mountainsides. The tongue was a vertical wall of ice that calved box-store-sized chunks into the bay.

Greenland was a blank spot on the map at the turn of the nine-

Justus Utuaq spent about half of every day wrestling the lark's foot to untangle the trace lines, usually with bare hands. "It's good to have something to do," he said.

teenth century, one of the last unexplored corners of the planet. As such, the island and both poles became the focus of adventurers and earth scientists, both looking to unlock secrets about how our planet works and to get their names into history books. You have to clear your mind of everything you know today—weather reports, computer models, satellite images—to understand where these explorers were coming from. Science in the early 1800s was still grappling with quandaries like what caused appendicitis, how to keep a silver nitrate photograph from fading, and the discovery of massive ice sheets on Antarctica. Ice to civilians was what you saw on the top of a mountain or on a pond in the winter. There was no common understanding of ice ages and, therefore, no comprehension of the planet's complete heat and water cycles, much less how the topography of the world came to be.

A young Norwegian zoologist named Fridtjof Nansen was fascinated by the study of Greenland, including a theory by Adolph Erik Nordenskiöld about a verdant oasis possibly located in the middle of the ice sheet.* Nordenskiöld attempted to cross the nearly 700-mile-wide island in 1883 to prove his theory. He made it 100 miles before turning around. Nansen had seen Greenland years before, as a researcher on a seal-hunting ship in 1882, and devised a plan to cross it. He was studying zoology then, gathering samples from the Denmark Straight. Oceans near the poles can freeze overnight given the right conditions, and when Nansen's ship froze in place, the young scientist asked the captain if he could walk across the ice to the coastline floating in the distance. It was around 25 miles away, across jagged icebergs, shattered pack ice, and patches of open water. The captain denied his request. A year later, Nansen began planning the crossing.

Nansen was not a typical geology student. He was an extrovert who craved adventure—tall and sinewy with a shock of blond hair that often stood straight up. Despite his flamboyant nature, he could be dark and brooding when alone. A passionate skier, he sometimes set out solo for days in the snowfields around Oslo. He did not seem to feel fear—or cold—the way most people did. Every adventure was a problem with a solution. The solution to a Greenland crossing would be seven-foot birch-and-oak skis—and modern telemark bindings—to keep his team on top of the snow on the way up the divide and sliding down the far side to make better time.

Six years after he first spotted Greenland's shores, Nansen returned with three Norwegians and two Laplanders, culled from forty applicants who volunteered for the expedition. All were expert skiers and

---

* In March 2021, scientists discovered fossilized plants in cores taken from beneath the Greenland ice sheet, suggesting that some kind of tundra ecosystem existed there in the last million years—and that the ice sheet is far more susceptible to melting than previously thought.

Fridtjof Nansen, photographed by Henry Van der Weyde in 1915.

athletes. One was a ship captain. Most inland expeditions at that point started from the west coast. Nansen's plan was to clear the difficult and dangerous task of rowing a boat through the pack ice in the east at the beginning of the trip, when the group still had energy and supplies, instead of at the end. They would then traverse to an established town on the west coast. The ship he hired to deliver the team immediately ran into ice off the coast of Greenland. Stuck in the ice, along with a dozen sealing ships, the crew waited for their chance to launch smaller boats they had packed and to make for the shadowy line on the horizon.

Six weeks after they left Norway, Nansen woke to find that the ice floe had delivered them to within ten miles of the coast. The crew launched two boats and made good time at first, rowing across open stretches of water and chopping through sea ice with axes. When they hit pack ice, they hauled the thick-hulled wooden boats up and over ice chunks the size of school buses. When Nansen's boat was pierced by ice that evening, they stopped to repair it and pitch a tent for the night. In the morning, Nansen realized that the ice floe had reversed direction and that they were headed out to sea. They were now thirty miles from the coast, and swells from the open ocean heaved them twenty feet in the air. That night, the floe cracked in half, just a few feet from the tent where the men were sleeping.

Waves crashed over their bivouac as the ice chunk shattered and shrank by the hour. They drifted farther out to sea until the fourth day, when a current slung them back toward the coast. For another week the crew rowed and navigated floes and currents, working ever closer to shore while drifting farther south. Eleven days after they set out, the boats nudged the firm, rocky shores of Greenland and the men rejoiced with hot chocolate and biscuits. It would take weeks to row north to their intended landing spot.

On August 9, the group packed five wooden toboggans fitted with steel skids that Nansen had helped design. The Greenland Ice Sheet does not slide gently into the water; it crumbles and buckles around the edges into hundred-foot ice cliffs and crevasses. Nansen found some of the smaller crevasses by falling into them, catching himself with outspread arms. The going was slow on steep terrain, especially hauling the two 100-pound sleds. They had to travel at night to avoid slushy, heavy snow. Two weeks into the crossing, thankfully beyond the crevasse zone, they had climbed from sea level to six thousand feet and were closing in on the divide.

The first snow to fall on the Greenland Ice Sheet accumulated nearly a million years before Nansen's foray on the ice. At the divide, the snow is more than two miles thick at an elevation of 10,551 feet.

Nansen and his crew spent half of the trip climbing up, and half coming down. The temperature got so cold as September approached that Nansen decided to shorten the journey by a hundred miles and headed to a different town farther south. The incline eased as the crew neared the summit of the ice sheet, but conditions hardened when the cold set in. Nansen was a constant cheerleader when storms pinned the crew down. They would huddle together for days, three to a sleeping bag made of reindeer pelts, praying that the tent poles would hold. One night, Nansen's thermometer dipped to minus fifty degrees Fahrenheit—at that time, the coldest land measurement ever taken in September.

When the snow hardened, the men could ski, and they made better time. The incline flattened closer to the divide, but early autumn storms pounded them every other day. Nansen made hot drinks and food for the crew: two pounds of biscuits and dried meat was the daily ration. Eventually, the angle pitched down to the west and the group picked up speed, fitting small sails to the toboggans to help push them along. On September 17, one of the men spotted a snow bunting flitting about, and soon after, they came to the crevasse zone on the western side of the island. The mountains and coastline of western Greenland lay before them. It would take another week to get off the ice and make it to the ocean. There the crew built a small reed boat from willow branches and the canvas tent, and paddled and dragged it to the village of Godthåb (today's Nuuk). Danish officials welcomed them there with cheers and news that the last ship for Denmark had sailed two months before. Their reward for crossing Greenland would be to spend the winter there.

It didn't take long for news of Nansen's success to circle the world. Nineteenth-century media was frenzied about exploration, and by the time Nansen and his crew returned to Denmark the following year, he was a national hero. He had proven the impossible and opened up a massive new territory. He had also single-handedly put the sport of skiing on the map. Skis were soon exported from Norway around the Northern Hemisphere, landing in far-off mountains from California's Sierras to the Italian Dolomites. His trek also opened the door to scientific work on the island, inspiring a long line of researchers who would ultimately come to understand the properties of ice and the ancient story it could tell us—as well as how the Greenland Ice Sheet alone could determine the future of practically every person on the planet.

The Farthest Shore

Ghosts of race day loomed over the track as Mikael casually took the lead near the end of Torsuut Tunoq Sound. Justus pushed his dogs to catch up, and the sled accelerated. The runners rumbled when we hit bare sea ice where the wind had blown the snow off. The air—not the sky but the air around us—was milky with ice crystals swirling off the track. It was hard to see more than a half mile, and I began to feel the first touch of cold penetrating my jackets.

Justus eventually overtook Mikael, who we later learned had been texting his girlfriend and didn't notice. Justus angled east toward a low pass where the snow was deeper and the dogs struggled and slowed. You get to know the cadence, hierarchy, and power—plus an intimate understanding of each animal's relationship with its alimentary canal—staring at the south end of a dozen dogs headed north for eight hours. The amount of diarrhea erupting before us suggested that at least half the dogs were not well. In fact, Justus told us, they were in prime shape. He had been training them for the East Greenland Championships for months. This would be the last week of training before they made another bid for the crown.

The dogs hauled in sequence, each one pulling hard for a minute or two, then easing off. The lead dog looked like a husky from a Jack London novel: tall and lithe, beautiful white-black coat. He flirted with his diminutive female partner. Not every dog was so lucky. Fights broke out every half hour and left blood on the trail. When we hit the ice foot between the inlet and land, the team strained against the harnesses, pulling us up a steep slope through knee-deep snow-drifts. The race track was gone; we were on our own now. Justus cracked a fifteen-foot whip alongside the dogs. The whip is made by peeling a spiral of skin off a dead harp seal the way you would peel an apple. It isn't used to hit dogs. Mushers snap it on the snow to encourage them to turn or pull. The sled lurched, then stopped, and Justus got off, ordering us to stay put and for the dogs to keep pulling.

We continued slowly, painfully up the steep incline—the dogs fought for every foot, stopping dead in snow drifts and getting a bit of speed near the top, where the powder hardened into a crust. The drop-off on the other side of the pass was steeper, longer, and completely nerve-racking as we picked up speed. "Four hundred vertical meters!" Justus yelled gleefully as all fifteen hundred pounds of us hurtled down the pitch at twenty miles an hour.

The sensation of accelerating downhill was familiar, but sitting in the back with no control felt like riding a roller coaster that had gone off the rails. The sled frame jerked from side to side as the runners hit ice bulges and ruts. Subzero wind blasted my cheeks and seeped through the zippers of my jackets. The dogs were no longer pulling; they were just trying to not get run over. At the bottom, we hit smooth hard-pack snow and glided to a stop. As if to make the scene even more sublime, the wind died. Clouds peeled back to reveal blue sky, and the temperature warmed fifteen degrees. We were in another inlet, a finger of sea ice in Aqartertuluk Bay that reached behind Taatsukajik Mountain. With no visible outlet to the sea, the bay looked more like a lake than the ocean. Low, rocky hills rose in every direction. The dogs kept their pace for the next few miles as we rounded the northern shore. Thirty minutes later, I spotted a sliver of white leading to the pack ice of Igterajip Ima Sound and the mountains of Erqiligarteq Island.

The night before, Spenceley had sneaked away from the dinner table—set with an outrageous Tex-Mex buffet—to show us his collection of antique Greenland expedition books. He laid down a small square of fabric before placing the books on the table, eyes wide, perhaps the same wonder he felt two decades ago on his first trip to the island still just as powerful. It seemed that he had been looking for something to wrap his life around when he first stumbled across the island. Ever since, he protected the island, its people, and their culture the way anyone would a living talisman. Within the wide and brilliant spectrum of that love was a particular fascination with historical accounts of young men who, like him, had undertaken ridiculously difficult and risky self-made adventures across it.

Grainy black-and-white images in the book showed two young men from England piloting a small pleasure boat around the coast, dipping into bays and fjords, donning climbing gear and exploring terrain that until then had been the sole domain of the Inuit. In the back of the book, Spenceley carefully unfolded what looked to be a waxed paper map of Kulusuk and the coast marked with ancient dogsled trading routes. He pointed a calloused finger at a dotted black line wending north from town through mountains and bays. "This is the route they hunted and traded on," he said in a whisper. "This is where you are going."

Justus now drove his dogs hard for another forty minutes before stopping at a small hunting shack to wait for the others. The shack was paneled with weathered wood. A piece of green nylon rope and seven nails served as a doorknob and lock. Inside, I found names and dates etched into the wood. A bottle of lamp oil, three red candles, and a scrub brush for dishes rested on a small platform. It was surprisingly cozy out of the wind, and I unwrapped a sandwich and drank from a thermos of hot tea that Theresa had packed.

Manterfield said that hunters did not use the hut much anymore. Years ago, the Inuit hunted for seal, reindeer, Arctic hare, eider duck, and polar bear. Now the hut was used as a spiritual jailhouse—where villagers possessed by evil spirits were interned for weeks or months to exorcise the demons. Danes converted most Greenlanders to Christianity in the 1700s, at the hands of zealous Lutheran priests and missionaries, but spirits of the old world still exist today. Pack ice can be smoothed by giants, and bears can be killed by a human taking the form of another bear and beating it in battle. Mountain spirits live in cracks in the mountains and run faster than caribou. They carry mirrorlike objects that reflect everything about a human being—your nature, your fears, your pride, and when you are going to die. The spirits want to be known; they kill anyone who doesn't tell others about them.

To keep the weather gods happy, rituals must be undertaken precisely. If a hunter kills a seal, he must pour water into the corner of its mouth immediately. If he hunts a whale, he must sleep with his wife the night before. The wife of the chief stays in bed during the hunt and must keep one foot on the floor. If the hunters are victorious, the wife will bring the whale a sip of cool water.

Animal parts hold great powers as amulets. The foot of a northern diver duck makes a person a better kayaker. Bear teeth and bits of dried flounder protect against attacks from other tribes. The skull of a raven or its talons guarantee a better share of meat from the hunt. Hunters can protect their heads by wearing a dried bee sewn into their hoods. A strip of salmon makes the stitching on their clothes stronger, and a water beetle strengthens the temples of their heads. A ptarmigan skull makes a hunter faster, and the head of a tern brings fishermen good luck.

If someone is lazy in their life, they are confined to a world just beneath the surface of the earth when they die—where they hang their heads in shame for eternity. The greatest hunters and shamans go to Aglermiut, a joyful place in the middle of the earth where they can leave their tent by spitting straight up and flying through the saliva. To get to one of these places without dying, you need to travel deep into a dream. Inuits hang on to the real world when they dream by gripping their corporal body's big toe. If they let go, they die.

Shamans were in charge of interpreting and enacting much of this magic. It was not an easy position to attain. Aspiring shamans have been drowned, starved, left out in the cold, and sometimes shot to open a path between them and the spirit world. They could be left without food for weeks or months or given only a sip of water between fasts. They could be dragged behind a sled or left in a snow hut for a month, where they were expected to give themselves over to the spirit world in a series of hallucinations.

The spirit jail, on Aqartertuluk Bay.

A sliver of snow ran lengthwise along a bedside table in the hut. Silver light pushed through a crack beneath the door. I heard yips and barks from Mikael and Mugu's dogs, who fought as they approached side by side. Manterfield was waiting next to the sled when I stepped back out of the shed. The wind howled, and bits of ice and snow pelted my goggles. Without six inches of high-tech layering, I would have frozen in fifteen minutes. Justus proved this theory wrong, hanging on to the rear handles barehanded, waiting impatiently for me to hop on the sled.

As I climbed onto a duffel bag, Justus scurried to the front of the sled as it took off. We were in deep snow again, and the dogs struggled. Whenever the runners hit a drift or a rock, the sled stopped and Justus hollered commands and cracked the whip. He was so accurate with it that he could whip a soda can off a person's head. (He did exactly this to Wendy on the last day.)

The dogs struggled for forty minutes up a long hill. Our pace was slower than walking speed, adding to the guilt I already felt reclining on the squishy duffel, watching the dogs toil. Back on the ice, the snow mixed with seawater, creating pale blue slush and slowing the sled nearly to a dead stop. The wind had picked up again, hitting forty miles an hour at times. With no trees or buildings to block it, it felt like it was pushing us backward. The gusts were bitter cold now, and even the unshakable Wendy seemed frozen in place, unmoving except to swing her arms, trying to get her circulation flowing.

The Shit Knife

Because Inuits lived so remotely and have never adopted a written language, few knew about or understood the civilization for millennia. Then, an eccentric self-made ethnographer-explorer took it on himself to circumambulate the land of eternal winter on a dogsled and interview the people living there. Knud Rasmussen was born in

Greenland in 1879—in Ilulissat (or what the Danes called Jakobshavn), a village set on the dark, frigid waters of Disko Bay on Greenland's west coast. His father was a Danish missionary. His mother was mixed Danish-Inuit. Rasmussen grew up speaking Danish and Greenlandic, listening to hunters' stories about tracking polar bears and walruses in the far north and joining town feasts when groups returned home with a narwhal or a polar bear. He started mushing dogsleds when he was ten years old and traveled with his father around the coast, christening churchgoers. A year before that, he and a friend hiked for miles east of their town, trying to win a cash prize offered by a local newspaper to the first person to spot the arrival of an intrepid explorer named Nansen, who was attempting to walk across the ice sheet.

Rasmussen's world was pulled out from under him and replaced with the neat, grid pattern of Denmark after his father was reassigned to a church there. The younger man initially struggled with school and the rigidity of Danish culture, but his charm and growing circle of artistic, bohemian friends soon made him a neighborhood celebrity. His mixed-race looks stood out, and he had a natural affinity for people, including a long list of heartbroken women. Rasmussen threw parties for any reason, at any hour, dancing and singing until dawn, with or without booze and tobacco to pass around. He was a free spirit and a budding Renaissance man who one day decided he would be a stage actor, the next an opera singer, then a journalist, and then, finally, an explorer.

To Rasmussen, the field of journalism was the study of his two passions: people and place. After college, he wrote a well-received travelogue about Iceland. He had traveled with a classmate from college, Ludvig Mylius-Erichsen, who invited him in 1902 to join a "literary expedition" to Greenland. Mylius-Erichsen planned to write about Greenland's people and geography. It was Rasmussen, though, who would end up spending the rest of his life on the island documenting Inuit culture, writing seminal works about the Inuit people, undertaking unthinkably ambitious expeditions, and earning the title "father of Eskimology."* Rasmussen picked up where he had left off as a young boy in Greenland, running dogs, hunting, and saving the expedition group on several occasions. During eight months of dog-sledding on Smith Sound, he interviewed dozens of Inuits, a culture largely unknown at that time. Two years later, he published two books about life in the polar north, both of which were well received, launching his literary career. He married a wealthy young Danish woman and moved into a home near Copenhagen. The marriage was made, however, with the understanding that he would be gone most of the time in Greenland, finishing the job that he had started.

Rasmussen laid the groundwork in 1910 for what would be five major expeditions in the Arctic, conscripting his friend Peter Freuchen to build a trading post and base for the missions with him. Their first stop on the west coast was at Rasmussen's uncle's house near Godhavn (today's Qeqertarsuaq); he was a successful trader who drank gin with them, showed off his marksmanship by shooting friends' tobacco pipes, and serenaded his guests with a violin (him) and an accordion (his daughter). After securing dogs from his uncle, Rasmussen made his next stop in a village called Uummannaq, on North Star Bay, at latitude 77. There they christened their new base Thule, the name of Pytheas's mythic origin of winter. Rasmussen's plan was to continue a trading business that Robert Peary had started during his quest to reach the North Pole, offering knives, guns, and tools—which could help Inuit families avoid starvation and compete with

* What separated Rasmussen to some degree from colonists and imperialist-minded explorers in Greenland, like Robert Peary, was his lack of interest in conquering, making history, or extracting resources. Rasmussen was born in Greenland and was part Inuit himself. Many of his explorations were merely an effort to answer childhood questions about the people he grew up with, the island, and the mysterious *tunu* in the north.

modern civilization as it encroached. The profit Rasmussen derived from selling the furs and tusks the Inuit people had traded for guns and other tools would finance the base and Rasmussen's expeditions.

Freuchen and Rasmussen assimilated quickly with the villagers, some of whom immediately moved in with the explorers in a prefabricated house brought from Denmark. Freuchen hunted caribou with locals to feed his dogs and learned to use a walrus harpoon.* Rasmussen and Freuchen discovered abandoned huts supposedly occupied by ghosts and watched Inuit children play naked in freezing meltwater lakes, grabbing salmon with their bare hands and eating them raw. On a hunting expedition to Ellesmere Island, the two men helped deliver an Inuit baby halfway across Smith Sound in the middle of a blizzard—standing around the sled to block the mother from a brutal Arctic wind. (Her husband made a slit in her pants so she didn't have to take them off.) Back at home, after Freuchen thanked a villager for a gift of meat, he was taught a lesson he often referred to: "Up in our country we are human!" the villager told him. "And since we are human we help each other. We don't like to hear anybody say thanks for that. If I get something today, you may get it tomorrow. Some men never kill anything because they are seldom lucky or they may not be able to run or row as fast as others. Therefore they would feel unhappy to have to be thankful to their fellows all the time. And it would not be fun for the big hunter to feel that other men were constantly humbled by him. Then his pleasure would die. Up here we say that by gifts one makes slaves, and by whips one makes dogs."

The ensuing missions were officially meant to conduct research in geology, botany, and biology of the north, along with Eskimo culture. In reality, they were complete ethnographic surveys of the

---

* The harpoons were tethered to a sea anchor, which was made from animal skin stretched across a frame to slow the beast down, and an inflated bladder to keep it afloat after it died.

Inuit people. Planning the expeditions was not so different from engineering a journey to the moon. They had to learn or devise new ways of dressing (animal furs), eating (raw seal meat, and a paste made from mashed caribou marrow and brains), and camping (often a snow cave or simply curled up behind a sled).

The first Thule expedition, in 1912, consisted of four sleds, four people, and fifty-four dogs. The mission was to map the northern coast of Greenland—and challenge Peary's claim that a channel divided "Peary Land" from the island—on a 620-mile dogsled adventure on which nearly everyone died.* The group was half starved and fatigued at the end of the first three-week leg. The members spent the next three months hunkered down in snow caves and pits, hunting musk ox, and healing from the journey. They documented everything they saw: sheer ice cliffs, northern lights, Polar Eskimos, and ancient settlement sites that made Rasmussen wonder if Inuit hunters had migrated to Greenland from the west. All four explorers returned, though only eight dogs survived.

Over the next seven years, Rasmussen set out on three more expeditions, mapping the north coast, supplying Roald Amundsen's effort to drift in the ice across the Arctic Ocean, and documenting Eskimo villages all over the island. He interviewed an old woman who remembered a time before such expeditions, before Peary brought guns and knives to Greenland on his search for the North

* Peary did not have a reputation for chivalry, or accuracy, in Greenland. His deep desire for fame and his imperialist tendencies often led him to exploit the local population, forcing Inuit women to work nonstop for days making reindeer sleeping bags and stitching clothing. He kidnapped an Inuit boy named Minik in 1897 and brought him home to New York to use him as part of his Arctic act—an extravagant Buffalo Bill–esque show he put on to raise money for future expeditions. One day he took Minik to the Natural History Museum to show him the new Eskimo display. Minik was horrified to see the skeleton of his grandfather there and asked if he could go home. Peary denied him for a decade, after which Minik finally made it home and settled down. Minik was one of the first to move in with Rasmussen at the Thule base.

Pole, when tribes hunted with harpoons and preserved whale and walrus meat by fermenting it. She divorced her husband during one especially cold and difficult winter—because he ate a bag of puppies without sharing them with her. On Rasmussen's way home from the woman's village, he and his team discovered a meteorite. A few days later, they found an octopus fossil ten million years old.

The fifth and last expedition was Rasmussen's swan song. He traveled twenty thousand miles over three and a half years by dogsled, and occasionally boat, from Greenland to Nome, Alaska. His mission had become increasingly focused on tracing the Inuit west, back to the land from which they came. Following their migration in reverse, he discovered tribes that still hunted with bows and arrows and used knives shaped from yellow flint. He found caribou antlers milled into harpoon shafts, bear shin bones used for harpoon heads, and seagull legs whittled into sewing needles. Visiting one tribe in Arctic Canada, he was told that they often hunted naked, in winter and summer, to be more stealthy in the woods.

Word of Rasmussen preceded him, and he was typically met by exuberant villagers who would often build a traditional dance house to receive him. (Rasmussen's desire to dance all night often outmatched theirs.) In the summer, he and his Inuit wife, Arn—it was common practice for explorers to take a second wife during a long journey—walked among nesting swans and geese on the shores of Disko Bay. Red patches of saxifrage grew on the grassy plains in the summer, where Rasmussen fished for salmon in wide, glacial-blue rivers. Ducks paddling in lakes and streams appeared like a mirage in the short summer season, a glimpse of the climate the rest of the world enjoyed.

He found Eskimos who still practiced the intricate art of bird skinning, crafting socks and underwear from tanned bird skins, and he discovered ancient sleds with walrus bone runners and handles made from frozen fish. One day in Arctic Canada, he found an old

Knud Rasmussen as a young man. In reading Rasmussen and Nansen's accounts, one similarity regularly stands out. There was seemingly no end to what they were exploring. The ice edge shifted; entire villages moved. Travel routes over fjords and bays froze, melted, then refroze overnight. (Bain News Service, part of George Grantham Bain Collection, Library of Congress, www.loc.gov/item/2014682626.)

man sleeping on a bed made of caribou ribs. The windows in the elder's igloo were made of frozen slabs of fresh water. Rasmussen dogsledded past ice drifts north of Illorsuit Strait; these drifts would one day end up in Baffin Bay, then Georges Bank off the coast of Maine via the Labrador Current. When he had to use a boat in the summer, he strung up the dogs to pull it from shore or an ice edge. He heard stories about a heartbroken bachelor who turned everyone in the village into stone. And two sisters who escaped persecution for stealing a caribou skin by becoming thunder and lightning. In another camp, villagers told of a mother who skinned her daughter and wore it to trick her son-in-law into bed. On Baffin Island, Rasmussen found a village that had recently lost five hundred of its six hundred residents to starvation because the caribou hadn't migrated.

The dangers Rasmussen, Freuchen, and the other team members encountered, and somehow eluded, are difficult to fathom. On one occasion in 1922, when Freuchen was trapped by a wicked winter storm near Repulse Bay, he wrote in his account of the journey that the temperature dropped to sixty below zero. The trail was obscured by wind and snow. Freuchen walked in front of his dogs, finding the way until the wind was so strong it became difficult to breathe. He dug a snow cave for himself and the dogs but was soon buried. Ice formed on the inside of the cave, making it impossible to dig out—so he defecated, shaped a knife out of his own excrement, and let it freeze solid. He dug himself out with the knife, but both of his feet were frozen and he couldn't walk. The dogs had run off by then, so he crawled for three hours to the Inuit camp he had come from. His friends, seeing that his toes were developing gangrene, cut a hole in their igloo to put the stinky, rotten flesh outside. When the skin and muscle on his toes began falling off and he could see the bones, Freuchen used a pair of pliers and a hammer to knock them off. He eventually convalesced and returned to the coast, only to be stranded on an ice floe that broke off from the shore while he and an Inuit family waited for a ship. Five days later they found land again, and he walked to a village on his stumps. Freuchen's leg would be amputated five years later, bringing his expedition days to a close.

Rasmussen's vision and energy held out until a few days before the seventh Thule expedition was set to begin. During one of his renowned feasts, he ate *kiviaq,* a fermented delicacy of rotting birds placed in a seal stomach and aged underground for a month, and got food poisoning. In his weakened state, he developed pneumonia. Rasmussen died in Copenhagen at the age of fifty-four. His gift to the Western world: twenty thousand Inuit artifacts, six thousand pages of writing, and the first record of the world's northernmost civilization.

"I Hope Then We Will Have Ice"

Our journey was painless compared with Rasmussen's, yet there was still some question whether we would reach the first camp before dark. Deep, wet snow continued along the inlet, miring the sled and dogs and slowing us to a crawl. An hour later, we turned a corner and hit some hard, wind-scoured ice. The runners growled, and we took off so fast that I worried about breaking an ankle on an ice chunk whizzing past. (Sitting sideways on the sled, our feet often dangled over the side.) We covered ten miles in the next hour, still charging into the gale, ghostly bergs raising their heads from the sea ice, then disappearing behind us. Justus and Manterfield chatted, like two men waiting for an elevator: what was the weather like, what were the ice conditions, how far north might we make it? The exposed ice was a good sign, Justus said. If it stayed cold and clear, we might make it to the remote *tunu*.

Thirty minutes later, Justus stopped the dogs near a five-story iceberg to look for seals. Mugu and Mikael followed him around the berg as Justus chipped holes with a five-foot ice pick similar to the one Lars Bianco had wielded in Kulusuk. The pick was like a multi-tool for the hunters. They used it to test the sea ice, to chip holes for ice fishing, as an ice axe for climbing, and, as a last resort, as a spear against a charging polar bear. Justus found deep water on the third try and chipped two more holes in a line about six feet apart. He then scooped ice and snow from each hole with his bare hands, cupped his hands around his eyes, breathed into the hole to melt any remaining crystals, and peered through the water to look for underwater ledges or obstructions. Satisfied with the location, he attached a lead line to the ice pick and used it as an underwater javelin, stringing a net between the three holes. The trap was simple, if not brutal. When a seal spotted the breathing hole, it would happily swim toward it, get its flippers tangled in the net, and drown.

I had mixed feelings about this feature of the expedition. One of Grey's favorite stuffed animals was a seal with giant blue eyes. I had bought it at a pharmacy while standing in line. I could see her cuddling the little creature as Justus strung the net, and I secretly hoped a seal didn't fall for it. It wasn't that Inuits treated seals poorly. After a seal is killed, hunters ritually remove the liver, close the incision to keep the blood in, and share the meat among fellow hunters. Seal skulls were considered the center of the creature's soul and were placed in a circle back at the village. Each skull was pointed in a different direction, so the souls could find their way back to a body to be reincarnated. This was the cycle all animals went through: birth, death, reincarnation. If it was disturbed, the Inuit people would starve.

Starvation was definitely off the table when we arrived at the first camp just before sunset. Stashed here was enough dehydrated food for several NASA missions. We found the hut above a steep embankment that Manterfield instructed us to walk around while the hunters drove their teams practically straight up the forty-foot wall of snow. The camp looked like a small Quonset hut and was outfitted with sleeping bags, plywood bunks, a picnic table, a stove, and cooking gear. Manterfield shifted into guide mode and melted snow for water while we helped the hunters dig snow anchors and hook up the dogs to chain tethers. The dogs howled until they were fed a couple of pounds of seal blubber, then they settled down for a nap. Back at the camp, our heroes clicked into gear: Scott and Wendy dug the latrine. Jason and Mayah thawed each other out with hot drinks and a sleeping bag. Manterfield turned on lights and an oil furnace and unpacked all the gear we would need, while I wandered around outside, stunned by the twilit sky and endless line of unnamed peaks reaching in every direction.

Justus stopped me as I headed up the side of the mountain the lodge was perched on. "Where are you going?" he asked. I said I was going up the mountain, and he smiled and fist-bumped me. Manterfield intercepted me a minute later and handed me a small plastic box filled with flares. "If you see a bear," he said. "Fire one of these at it." It was a pack of five; three were gone.

I scrambled up the ridge for a half hour, postholing through deep snow and hopping from rock to rock. There were no bears in sight. Given the typical predator–prey relationship, I figured I wouldn't see one until it was too late. Halfway up the ridge, I sat down and leaned back against a warm boulder. The sun was still bright and strong. The wind had slowed to a somewhat normal speed. Two massive summits dominated the horizon to the north. I could see the start of a distant range to the northwest and a long patch of open ocean. Monolithic white towers of ice in the east, broken off from Greenland's glaciers, stood frozen in the sea.

The sky faded to blue as the storm dissipated. The sea ice to the east held its blue hue as well, as did the clouds and the swirling snow. The ground was rubble, broken stone sprinkled with windblown snow. The landscape looked so much like moon-landing pictures I had seen, that with my down suit and goggles on, I felt as if I had landed on another planet. It was another alien scene, like the one I had seen from Washington Pass, where I had started this journey: a glimpse of the planet and its story, separate from ours. I thought about microbes frozen into comets and meteors raining down on the planet over the last few billion years. Many of them hit right here. Signs of life recently found in Greenland's bedrock date back three and a half billion years.

Night set in, and the temperature plummeted. I headed back to the relative warmth of the hut, where the hunters sat in aluminum folding chairs around a diesel furnace. The white, insulated material of the roof arched over their heads like an adobe chapel. Parkas and snow pants hung from two green cords strung lengthwise from the door to the sleeping area. Behind that was a steel picnic table stacked with gear, and on the opposite side a pine shelf with a half dozen clear plastic storage boxes filled with supplies.

Manterfield offered us a selection of exotic meals in foil pouches—Orzo Pasta Bolognese, Chili Con Carne, Spicy Pork Noodles, Posh Pork and Beans—then poured boiling water into each. We sat around the furnace for an hour, working our way through the incredibly dense servings. Justus showed me pictures of his answer to melting sea ice: 1,500-foot-long blue-and-black fishing lines tied with jigs he used to catch halibut from his boat. There were five hundred hooks on each line. He also fished for Arctic char and, now for the first time,

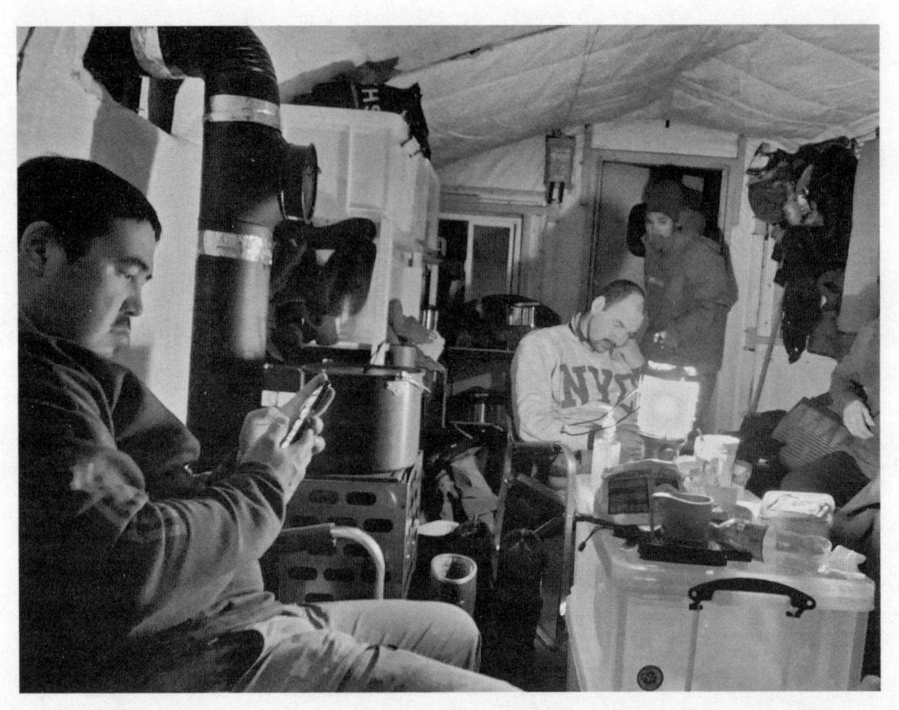

(From left) Justus, Mugu, and Mikael checking Facebook at the first camp.

cod and mackerel, as they migrated north to find colder water. When the season is right, he harpoons narwhals as his ancestors have done for thousands of years. He showed me a picture of the fruits of his labor: a cutting board in his kitchen heaped with narwhal mattak—two-inch blubber cubes—dried cod sticks, and a sliced apple.

He showed me another photo of him on his dogsled next to a giant swath of open water. "Temperature here the same for a thousand years," he said. "Fifty years up, fifty years down." He looked at his watch to see what year it was. "In 2024, the temperature will go down again," he said. "I hope then we will have ice."

17

Nancy Pelosi Goes to Swiss Camp

A few months before heading to Greenland, I visited one of the founders of Greenlandic glaciology, who was also based in Zurich. The glaciology network is a small one. Every scientist I spoke with about Greenland pointed me to Konrad Steffen, or "Koni," as colleagues, heads of state, senators, activists, billionaires, and, eventually, even I called him.

It took six months to track him down. Emails were exchanged, phone messages were left. Dr. Steffen never seemed to be in the same place for more than a week: New York, Bhutan, Norway, Africa, Antarctica. He was a lead author of the 2019 UN report on the cryosphere and was with Al Gore in South Africa the day before I was supposed to travel to Switzerland. A last-minute Hail Mary email from me confirmed that we would overlap in Zurich for about twelve hours. He sent me an address and said he could spare an hour. I stepped off a red-eye at seven in the morning and practically sleepwalked to a tiny and extremely efficient car rental agency. I loaded my gear into an equally efficient subcompact car and drove ten miles across town to the Swiss Federal Institute for Forest, Snow and Landscape Research (WSL).

The grounds of the institute at Zürcherstrasse 111 were lush and green, even in late January. Grass grew around a collection of blockish gray buildings. The structures were more glass than steel, with slate gray steel trimming the windows. Gray is the color of efficiency. It mimicked the gray winter skies, the unblemished gray asphalt roadways where scientists drove gray electric cars and pedaled bikes that they left in a small, gray parking lot without bothering to lock them up. The roof drains, which collect and redistribute water to landscaping and lawns, were gray as well, but the conifers growing outside the front entrance were an almost artificial shade of green, their branches practically growing down to ground level. A short path connected the campus to a grocery store; in the other direction were the tall glass doors of the entrance, behind which the hale, rangy figure of Koni Steffen walked up the stairs toward me.

Steffen's hand was outstretched before I even got through the door. A wide grin stretched across his face. He was sixty-eight years old and looked like a tall, thin mix between Fridtjof Nansen and Kris Kristofferson—white hair, a trim white beard, and the freckled, weathered skin of a seasoned explorer. His hand looked like something you'd want to grab if you were slipping into a crevasse or snow-blind on an ice sheet. He seemed like the kind of guy—cheery, strong, confident, smart—who could maybe help save us from the equally frightening threat of winter's end.

Steffen's office at WSL looked more like a penthouse suite than a science lab. Four thick rafters angled through the space at thirty degrees. Oversize glass windows lit the room. A small palm tree grew in the corner, and photographs of ice formations hung above three leather lounge chairs. Bookcases held studies, dissertations, and trade publications. A row of lab coats and blazers hung from a stainless steel coat rack.

Steffen was not one of those scientists who were uncomfortable in business attire. His father was a successful fashion designer, and Koni had inherited his easy aesthetic. He wore tailored slacks and a pin-striped button-up shirt with a French collar. His gray hair swept back from his forehead in two breaking waves. He had the sparkling blue eyes and wide smile of a teenager. You might think he was a dandy from a distance, but you would be wrong. This is a man who, on his first trip to Greenland, rolled a snowmobile five hundred feet down a hill, suffered a compound tibia-fibula fracture, set the bones himself, and managed not to bleed or freeze to death until his colleague began wondering where he was and miraculously found him a day later. There were no cell phones or satellite phones back then. "Thank God I remembered not to smoke a cigarette," he recalled. "It opens your veins." (Another survival story: One night he took his wife on a romantic camping trip on the ice sheet—sans rifle—to get away from the crew, and a polar bear tried to break into their tent in the middle of the night. They knocked it out by feeding it meat laced with morphine from the first aid kit.)

Steffen's journey through the cryosphere is as well documented as it is prolific. He has more than fifteen thousand academic citations stemming from forty years of research—and was the first to sound the alarm that Greenland's ice was vanishing and that its disappearance would affect the world. He did not want to be a climate scientist as a teenager in Zurich. Like Nansen and Rasmussen, he wanted to be an actor, but his parents convinced him to get a day job first, so he pursued electrical engineering in university. He excelled at math and physics and looked for ways to connect them to environmental studies, mostly so he could work outdoors. This was the 1970s, when climate science was called physical climatology, which he declared as his major. He eventually became a climate modeler.

The University of Colorado was making great strides in climatology then and offered Steffen a fellowship and then a professorship at the Cooperative Institute for Research in Environmental Sciences. The program, which he ultimately directed, discovered, among other things, a giant hole in the ozone layer. For thirty years Steffen worked with NSF, NOAA, and NASA developing satellites equipped with laser measuring devices.

Steffen also founded Swiss Camp, a lab that he set up in 1990 on Greenland's central ice sheet and that became a mainstay in the study of the cryosphere. His original destination was Tibet, but the Chinese government tried to charge his group exorbitant fees. Swiss frugality prevailed, and he deferred. The substitute destination was fifty miles east of the Jakobshavn Glacier near Disko Bay, not far from Rasmussen's hometown. Steffen and his students built a platform and installed three Quonset-style tents with eight-inch-thick insulated walls. Six snowmobiles, a fuel depot, a six-person sauna, and a surprisingly elaborate kitchen were soon added. Steffen stocked the kitchen with cases of his favorite wine, fresh fish from Scandinavia, and, occasionally, lobsters from the Maine coast. In pictures he shared of visitors at the camp, I saw several researchers I had interviewed for this book. Other distinguished guests included Al Gore, Nancy Pelosi, Anderson Cooper, and the president of Switzerland. (Steffen said that Gore gave the best toast at the reception for Steffen's second marriage.) When Switzerland hired Steffen back from Colorado, the Swiss president had to have parliament pass regulations to make an exception to the retirement age, which is sixty-five.

The celebrities were not there for Steffen's renowned culinary skills. They were there to talk about a great catastrophe—a biblical flood from rising seas. The revelation that the oceans are going to overtake coastlines around the globe is by now well known. More than 680 million people in low-lying coastal zones—plus 62 million living in small island nations like Tuvalu, the Maldives, and Kiribati—will be displaced in coming decades. The big picture is even scarier. Some 250 million people live on land less than three feet above the current sea level.* And 2.5 *billion* souls live within sixty miles of a coastline and could be on the

* The South Pacific is seeing nearly a half inch of sea level rise per year. Eight islands there have already been submerged. By the end of the century, forty-eight more are likely to be gone.

run from flooding in this century or the next. The fact that melting and calving ice is the cause of the flood is not as well known. Nor was what Steffen discovered on Greenland. The island was contributing more to sea level rise than were Antarctica and all the world's glaciers combined. In our list of global tipping points, Greenland held a place of its own. If the ridiculously rapid pace of the ice sheet melting and calving into the sea continued—imagine dropping handfuls of ice into a glass of water—the world's oceans would rise twenty-four feet.

Over the last few hundred million years, a balancing act has played out between land ice and the oceans. When the atmosphere warmed, ice at the poles melted into the ocean and sea levels rose hundreds of feet.* When the atmosphere was cold, water froze in ice sheets and glaciers, lowering sea levels again. Since the last Ice Age twenty thousand years ago, sea levels have already risen four hundred feet.

This synchrony is not exclusive to Earth; nor is the presence of abundant water. Water is everywhere in the universe. When stars die, the supernova that follows bonds hydrogen and oxygen, forming water, which makes its way onto planets, dust particles, meteors, and comets. Astronomers have spotted giant clouds of water hovering around black holes. In our solar system, water in vapor and solid form can be found on most planets—such as Mars, which, like Earth, is capped with icy poles and glaciers. Without the thick blanket of greenhouse gases that humans have draped over the atmosphere, we probably wouldn't have seen significant sea level rise for centuries. Melt from the Ice Age would have continued, unnoticeably, but our burning of fossil fuels has doubled the pace.†

Sea levels rose up to 15 feet per century shortly after the last Ice

* A third of sea level rise is attributed to seawater expansion as it warms.

† Since 1990, the rate of sea level rise has nearly tripled. Average sea levels have already risen more than eight inches since 1880; three of those inches happened in the last twenty-five years. (Sea level hadn't changed at all during the previous two thousand years.)

Age. Estimates for 2100 range from 16 inches to 8 feet. If both Antarctica and Greenland completely melt, oceans would rise 210 feet, inundating the coast of every landmass on the planet.* In the same way that humans can't feel one degree of climate change on their skin, the thought of 4 feet of sea level rise in our lifetime does not do the coming cataclysm justice. The creep of meltwater is not the only threat. It is the compounding effect of increasingly powerful storms, waves, and high tides that have already subjected 171 million people around the world to dangerous flooding. Average storm surge heights will rise by 10 feet by the end of the century, while the amount of endangered coastline grows by a third—including major portions of Maryland, Virginia, and North Carolina. A NOAA study cited a five-fold increase in Gulf Coast and Eastern Seaboard flooding in the last twenty years. If ice melt continues accelerating on its current path, we could be looking at 20 to 30 feet of sea level rise by 2200, which could move the entire US East Coast back to the I-95 corridor.

The situation is so dire that some of the largest geoengineering projects on the planet are aimed at stopping it: like a proposal to build floating seawalls beneath Jakobshavn Glacier—and Thwaites Glacier, where Seth Campbell was working—to block warm-water currents from melting ice from below. Other projects aimed at stopping melting at the poles include pouring tiny glass beads over exposed earth in Antarctica to reflect sunlight the way snow does and installing millions of wind-powered pumps to spread seawater over the continent so it will freeze and do the same. One of the more popular, and dangerous, schemes involves releasing calcium carbonate into earth's stratosphere to reflect solar radiation and cool the

* The gravitational pull of these ice sheets draws water to them. When they melt, the field diminishes, releasing water to the opposite side of the world. In this way, Jakarta feels the brunt of Greenland melt and New York City feels the effects of Antarctic melt. Recent melting on the poles has been so great it has altered the planet's rotational axis.

planet. (Some fear that doing so would open a hole in the ozone layer.) A more promising technology in Iceland is capturing carbon dioxide from the air and injecting it into basalt rock, in a way replicating and accelerating nature's own process.

In 2005, Hurricane Katrina created a storm surge of between ten and twenty-eight feet around New Orleans, razing buildings and killing more than eighteen hundred people. In 2012, Superstorm Sandy delivered a nine-foot storm surge to the New York–New Jersey coast, flooding eighty-eight thousand buildings, killing forty-four people, and causing $19 billion in damages. I was living in Brooklyn when Sandy hit. My office was in an old warehouse in the Red Hook neighborhood, about forty feet from a seawall where longshoremen once stacked bricks onto schooners. When I packed up my valuables and walked to my bike the evening before the storm hit, I found a foot of water in the parking lot. Side streets connecting to Buttermilk Channel and New York Harbor were already two feet under. Hours later, the city's pumps failed and subway tunnels flooded, causing the most extensive damage to the system since it was built more than a century ago.

I rode home on the only dry street I could find and still had to cut through a foot of water. We lived in Bedford-Stuyvesant then, which is set on what used to be rolling farmland in the middle of the borough, well above sea level. We never lost power or even internet service, so Sara and I were emboldened to head out that night to see the storm. The wind was like something solid, like a plywood wall pushing on your torso. We leaned forward into it, ducking behind buildings along the East River to see the Williamsburg Bridge swinging wildly and the wind knock the waves flat on the river's surface. Sara shot a video of the Con-Ed transformer on 14th Street exploding, plunging 230,000 people south of 39th Street into darkness for days. Con-Ed had prepared for a ten- to twelve-foot high tide. The fourteen-foot tide was more than its power stations could handle.

My graduate school professor lived on East 12th Street, and Sara and I rode our bikes over the bridge a couple of days later to check on him and his wife. People stood along First Avenue with pumps and buckets, trying to save their apartments. On Avenue C, trash, seaweed, foam, and other flotsam lay in the middle of the road. Just south of the intersection with 12th Street, the nexus of skinny jeans, tattoo parlors, and eight-dollar oat milk lattes—a good quarter mile from the East River—a thirty-foot piece of driftwood lay across the street.

Plans to save coastal cities such as Manhattan and Shanghai from sea level rise depend on engineering feats—like the $800 million Big U, a sea wall once designed to run around the lower part of Manhattan. But even those plans need an accurate timeline for how quickly the ice will melt—information that is pretty much Koni Steffen's job. Steffen maintains one of the oldest ice-melt data sets in the world, with thirty years of wind speed, temperature, solar radiation, and melt measurements from the Greenland Ice Sheet. His electrical engineering skills came in handy as he designed and built all of his own remote gauges. (He also helped build the time-lapse camera boxes that James Balog used to film his ode to the melting cryosphere, *Chasing Ice*.) With classified NASA clearance for high-definition, fifty-centimeter-resolution satellite imagery piped straight to his office in Zurich, he could see precise conditions on the ice sheet with a few clicks of his mouse.

I looked over his shoulder at a few data sets, then a comparison of summertime temperatures between 1864 and 1990 with those between 1991 and 2018. A black line rose gradually, then sharply, from left to right across the screen.* Steffen put on a pair of tortoiseshell reading

---

* For a big-picture, deep-time comparison, note that temperatures have warmed eighteen degrees Fahrenheit since the end of the last Ice Age twenty thousand years ago—during which time sea levels concurrently rose 425 feet. Considering these enormous temperature and sea level changes in a geological time frame, we're the ones caught off guard, not the planet.

Koni Steffen, the father of Greenland ice data, in his office in Zurich. "Anderson Cooper came to Swiss Camp," Dr. Steffen said, "and we did his show, 360°, from the camp. I took two beacons—these are our satellite transmitters—so we had 560k transmission and he could do a live broadcast. But he was high maintenance because these people are not used to the cold. He said he didn't change his clothes for three days, even when he slept, because it was too cold."

glasses and leaned toward the monitor, hovering his long fingers over the keyboard as he explained the chart. There was some pride in his voice at what he and his team had accomplished, but also discernible disbelief. He opened another chart and gazed at it. It was titled "Mass Balance at Swiss Camp: 1990–2018." Steffen's devices had calculated the plot points; the lines represented entire decades of his life—dangerous days avoiding crevasses, sixty-below-zero nights fixing satellite links, storms that ripped off pieces of Swiss Camp and flung them onto the ice sheet. Two textbooks held up one of his monitors; a simple calculator sat on the desk. Steffen ran his middle finger down the sloping mass-balance line as he described the loss of ice on the island: 3.8 trillion tons between 1992 and 2018. In the summer of 2019, extreme heat melted 600 billion tons of ice there, raising—in just two months—worldwide sea levels by 0.09 inches, or 40 percent of total sea level rise in 2019.*

Unlike Antarctica, where melt rates depend on the stability of floating ice shelves, the issue in Greenland includes surface melt. Warmer air temperatures melt the ice from above, then meltwater percolates through the ice to the sliding bed, lubricating its base and quickening its descent into the ocean. A spike in high-pressure systems over the island, likely due to an unstable jet stream caused by climate change, worsens surface melt with more cloud-free days. An estimated hundred billion tons less snowfall this decade, compared with the 1980–1999 average, has stunted glacial growth as well. Spring snow cover in June declined by more than 13 percent per decade from 1967 to 2018.

Much of this data comes from NASA's ice-monitoring satellites, ICESat and ICESat-2, the latter launched in 2018 with laser altimeters, the most accurate measuring tool ever deployed into space to measure the cryosphere. ICESat-2 recently took a deeper look into ice loss and found that between 2003 and 2019, Greenland and Antarctica's ice loss alone accounted for more than a half inch of sea level rise. "If you think about history, what is one or two hundred years?" Steffen asked. "Civilization evolved over a few thousand years at a very stable sea level with ten to fifteen centimeters of variability."

A rise of thirty feet would mean drowned cities and coastlines, but

* Scientists like to recite certain analogies to help people visualize size. For example, one billion tons of ice would fill four hundred thousand Olympic-sized swimming pools. And five thousand gigatons—the amount Greenland and Antarctica lost between 2003 and 2019—would fill Lake Michigan.

the more pressing question is where will those people go? And what kind of political and economic instability will follow? The current Syrian civil war was the first modern climate war, Steffen said, brought on by lack of water, high temperatures, failing crops, a spike in basic food costs, and all the tumult that has followed, including mass migration. It was a drop in the bucket compared with the upheaval to come, he said.

"Asia has all the megacities, and they are all at sea level," he said. "The area behind them is already heavily populated. These people cannot just move back. They move away, or they have to be moved away. One meter of sea level rise globally means—locally, with high tide and on-slope winds, maybe two to two and a half meters. It will close the cities, including New York. So where are the people going to go? Switzerland has eight million people, and if something like ten times the Swiss population comes west, I'm sorry, we are overrun."

Steffen knew firsthand what this radical transformation means. His biggest challenge when I interviewed him was trying to keep Swiss Camp itself from being swallowed by the ice. The elevation of the camp site has dropped by around forty feet since 1990, exposing its structural supports, tearing apart the platform, and destroying the much-loved six-person sauna Steffen had engineered. Crevasses have opened up around the base, making travel there by plane and even on foot impossible. "My son is a researcher, like me," he said. "We played baseball last year, and I almost lost him in a crevasse. He just was able to hang on by his arms until we pulled him out. It was a deep one. We started probing around and saw crevasses all over. We had to put in fixed lines, and researchers had to wear climbing gear every day. NSF made me report in daily until they said they were sending in a helicopter the next day to evacuate us. I wanted to finish my work, but they said no. The helicopters came in and we left."

We walked the immaculate halls of WSL after the interview and got lunch in the university cafeteria. It would not be an exaggeration or overly sentimental to say that I felt a grandfatherly connection to this man. He was all-knowing in his field but also self-effacing, generous, and irreverent. He held doors for me, talked me slowly through the complex data that he was collecting. He seemed grateful, blessed even, that his work not only had been supported but had also elevated him to a kind of celebrity. A wake of ogling students eagerly parted the way as we walked to the cafeteria. "This is the typical meal of the Swiss," he said next to a stack of gray plastic lunch trays. Option one was a mound of tuna with six potatoes and a hunk of cheese. Option two was vegetarian, which replaced the tuna with cottage cheese.

We sat in the corner with two younger colleagues. The conversation immediately turned to plans to get back to Greenland to see if Swiss Camp was still there. One of the scientists at the table said that Steffen's son was coming along after all. "Good news!" Steffen exclaimed. "NSF and NASA are funding us again. More food and good wine!" I was so caught up in the moment, and the benevolent spirit of the man, that I let out a spontaneous "All right!" as if I would be popping tops and digging into a steaming pot of Maine lobsters at Swiss Camp in a couple of months. I got the feeling that Koni would probably have invited me if I'd asked, but I was already late for my flight and bid a quick farewell instead.

Traffic along the shores of Lake Zurich raced by at seventy miles an hour. Just a few miles away were the ancient lake settlement sites I'd seen in Bern, where the first hominids settled in Europe and stared at the icy peaks with fear. They would have been living at the edge of the ice then, subjected to glacial outburst floods as the great rivers of ice retreated northward, receding back over the continent and filling the seven seas. The site was a bizarre confluence of time, bookends on the beginning and perhaps the coming end of human civilization's arc. Winter was the common thread stitched through our evolution, a great equalizer of both climate and human ambition.

Months later, I was working on this book at my desk in Brooklyn when I got word that Greenland had taken another great explorer. It was summer now. The cobwebs in the studio were still growing, but the cold draft was gone. Ivy tapped the window; an old fan hummed. Koni Steffen had walked into one of the crevasses he had told me about at Swiss Camp. Colleagues said he most likely drowned in water at the bottom of the fissure. His son was working at a study site nearby. Al Gore tweeted his sympathies, as did many leading voices at the United Nations, IPCC, NASA, NOAA, and WSL. I read tributes and accounts for the rest of the day, surprised at the grief I felt for a man I had met for less than three hours. It seemed an impossible end for someone so comfortable on the ice, but it also symbolized how quickly that world was changing. Baselines were vanishing; the impossible was now normal. After informing the world of widespread melting on the Greenland Ice Sheet, Steffen himself became a victim of climate change.

The Penumbra of the Planet

The air was still the second day of the dogsled trip, and the sun was out. The dogs lay curled up in the snow, heads tucked into their haunches. Bands of reddish rock cut through the snowfields on a small subrange of mountains to the east. The sun was already a few degrees above the peaks, hovering and milky, lighting but not yet warming the Arctic landscape. As if on cue, one of Mugu's dogs climbed a large boulder and, silhouetted by the postdawn solar explosion, arched its back, raised its chin to the sky, and let out a chilling howl.

I'd seen this before, in dog-eared copies of Jack London's *White Fang* and *Call of the Wild*. I read each book a half dozen times as a kid, lying prone in my twin bed, Hootie the Owl flannel sleeping bag pulled up to my chin, oversize steel-frame reading glasses perched on my nose. I didn't learn about the wild in school; I didn't learn it in church. As a ten-year-old on the coast of Maine, I was still, and would be for some time, trying to figure out what everyone in the world did.

During my adolescent years, I was pretty sure that I and everyone else would eventually end up in Alaska, lost in a storm and dependent on the loyalty and social hierarchy of a dog team to get me out.

This is the place the Arctic holds in our minds, going back to the Greeks and Pytheas and Aristotle. Cold-giving Thule is the last foothold of winter, sitting on the top of the world, a final, sacred symbol of the dark season. There is order in the Arctic's singular white plane, its ever-present view of earth's curvature, its simple food chain. Endless light follows endless dark. Loneliness, starvation, and exposure

"He had learned well the law of club and fang, and he never forwent an advantage or drew back from a foe he had started on the way to death. He had lessoned from Spitz and from the chief fighting dogs of the police and mail and knew there was no middle course. He must master or be mastered, while to show mercy was a weakness. Mercy did not exist in the primordial life. It was misunderstood for fear, and such misunderstandings made for death. Kill or be killed, eat or be eaten—that was the law, and this mandate, down out of the depths of time, he obeyed."—Jack London, The Call of the Wild

exist alongside beauty, quiet, and solitude. There are no trees or bogs blocking your way. You can always see where you are going.

Back in the hut, the crew was humming with caffeine-fueled conversation, apparel layering, and full-on psychotic fidgeting by several guests who had flirted with hypothermia the day before. Manterfield clinked his coffee mug and laid out a plan for the day. We would follow the Igterajip Ima Sound south around Igterajiik Island, then turn north into the blinding white field of Naportuit Nuuat Sound—an east-facing inlet that meets Sermiligaaq Sound and the calving faces of the Karale and Knud Rasmussen Glaciers. There we would find the tallest icebergs to spot seals from and possibly even drop a line in open water to fish for halibut.

An hour later we were off again, six flatlanders chasing three dozen dogs and their mushers down the hill, then south back the way we had come the night before. The dogs ran hard all morning and got tangled every thirty minutes. The Gordian knot they created was a thing of nightmares, a wad of nylon twenty inches in diameter, all worked out by Justus and the trusty lark's foot while underway.

The storm had wiped the ice clean, and we cruised at ten miles an hour all morning. The dogs intuitively knew the way, tracking the headlands, following inlets, cutting around open water and ice bulges that would have flipped the sled. In Naportuit Nuuat Sound, we cruised around twenty-story, sapphire blue icebergs cut with arches, knife-edge spines, and swooping flanks. Two hours in, Justus stopped at a long rocky ridge on the northeast coast of Erqiligarteq Island. Hunters used the ridge to spot seals and, we soon found, to get a cell phone signal. We scanned the horizon for little, seal-shaped dots—the idea of killing a cute seal was growing on the group—while Justus scrolled through his Facebook page and talked to a friend about a new boat. Frozen ocean ice braced the eastern shore; jumbled pack ice drifted by farther south. All around the bay, sharp granite peaks gave way to ridgelines running straight to the sea.

Manterfield pointed east and said that Iceland was just over the horizon. Three pyramidal peaks to the north had been landmarks for Icelandic sailors for thousands of years, he said. The easternmost summit was named for Erik the Red. They were all mentioned in Icelandic sagas as navigational aids for Norse sailors. After a lovely charcuterie laid out by Manterfield, we headed out again, angling north toward the fishing village of Sermiligaaq, and *tunu* beyond that. Justus guided the dogs through narrow passages between the icebergs, sometimes riding just a few feet from a 200-foot sheer ice face and occasionally stopping to hike an iceberg and look for a seal. I followed him up one, much to Manterfield's chagrin. Justus was thrilled and chipped stairs into the ice. Halfway up, I saw why he had done so. The vertical ice luge below us dropped straight into the Greenland Sea. We stood side by side at the top, searching for seals sunning themselves or coming up for a breath of fresh Arctic air. There was nothing but white, though, so we headed back to the sleds and took off again.

Two hours later we pulled into camp, which consisted of a red shack the size of a small bathroom. There were four bunks inside, three of which, we were told, would be occupied by the hunters. They needed their sleep and, hilariously, didn't like to sleep in the cold. This would be our winter camping experience, Manterfield said, in tents and subzero sleeping bags that he pulled from a bin a few minutes later. Each couple would share a tent; Manterfield would shack up with me. Mayah looked confused. Wendy and Scott did what stoic Northeasterners do when faced with an existential problem. They began working furiously—laying out a suitable tent site, digging a twenty-square-foot platform, leveling and smoothing said platform, packing down a vestibule area, laying out a tarp, then the tent, then the fly, beneath which they created a comfortable down cocoon to warm them in the cold black night.

We all followed suit, quietly, earnestly, while the three hunters anchored their dog chains in a line behind us, ostensibly to discourage

Justus Utuaq chipping steps into an iceberg on our way up to spot seals.

polar bears from wandering in. It could be said that at least several of us felt the proximity of death at that moment. It would not come in a flurry of bear teeth and paws, but with a slow lowering of body temperature. Sips from my flask helped Jason and Mayah pull out of a tailspin, as did the warmth of the shack and a lively dinner.

The hunters were already in their sleeping bags, and we sat at the foot of their beds, sharing stories about the day as we ate steaming Chili Con Carne and Posh Pork and Beans. We'd found a few more breathing holes, but no seals. The hunters didn't seem concerned. The hunting aspect of the journey lent some authenticity to it, but it was clear that keeping us alive and entertained was the primary objective. I retold some of the adventures of Rasmussen and Freuchen, including one passage about Rasmussen crammed into a similarly small shelter in an Inuit encampment. After falling asleep, he heard screams from a nearby tent—inside which he found a dozen Eskimos laughing and singing, mostly naked, in a giant pile. (Rasmussen joined the party for most of the night.)

The hunters were sound asleep by the time we headed out to the tents. The northern lights were supposed to be out that night. Mayah and Jason had never seen them, and the anticipation of the spectral light show was a welcome distraction. I climbed into my sleeping bag and unzipped the door of the tent to watch, just as a braided rope of green light reached from the northeastern horizon to the southwest. I'd seen northern lights in Maine, but they were cloudlike, with no defined shape. The light-rope was something else, a communication of some kind, a wand from another world. It wavered back and forth, like a line dragging behind a boat. After a few minutes it whipped ninety degrees to the north, then a few minutes later ninety degrees back.

Inuits believe the northern lights are an aid to shamans who can beckon the lights closer or spit at them and make them meld together. Light is magic here. Half a sun dog indicates that a death is coming. If a rainbow has steep sides, it means good fortune is on the way. If the arc is flat, a disaster will ensue. Thunder arises from singing thunderbirds that shoot lightning from their beaks, and falling meteors are excrement from the stars. Shamans are filled with light from spirits that enter through their navel and live in the breast cavity. They are bright inside; everyone else is dark so that the spirits do not notice them.

It all seemed possible, watching the show. What else could explain this? Gusts of electrons and protons spinning off the sun, blowing through space like a rain squall, colliding with oxygen and nitrogen in the atmosphere. The earth's magnetic field shields most of the planet from the barrage. But the field is weak in the north, and the particles get through. That's why Jim Koehler's IceCube project was stationed in the Arctic. Just like neutrinos, every collision creates a flash of light, meaning what we were now seeing overhead were seams in the planet's magnetic field, a penumbra of our planet, a celestial hairnet.

I watched the show until the freezing air—minus twenty degrees Fahrenheit at that point—burned my face. My eyelashes were frosted over and the thin, nylon sleeping bag exterior crackled as I moved. It is a strange sensation to feel your cheeks freeze while the rest of your body is snug and warm. I heard Mayah and Jason talking and Scott clicking long-exposure photos of the scene. The hunters snored in the little shack, and Manterfield, stressed out, was pawing through a storage chest behind the shack to make sure we had enough food and supplies to make it to *tunu*.

Green-blue light from the sky made the tent glow. This was the un-night of the Arctic. We were exposed to everything all at once perched at the top of the world, seemingly a few yards from space—the weather, the stars, the sun, the moon. It was part of what I loved about winter, the rawness of it, its irrefutability, how it took some skill to survive. It was pure nature with few people to spoil it. If you mastered it as Rasmussen did, you could have half the world to yourself all winter long.

I went full submarine into my sleeping bag to escape the cold, closing the cinch above my head. It took a while to fall asleep. I thought about Nansen in his deerskin sleeping bag, huddled for days, sometimes a week. How many explorers on this island never woke up from a night like that? How did Rasmussen survive for days in a snow cave with no food on the first Thule expedition? I nestled deeper into the bag, perfectly warm but still aware of the deadly cold a few inches away. This was a transition zone that Inuit shamans referenced—ice turns to ocean, life turns to death. I did not see a sun dog that day. I would not be visiting the moon or the stars in the land of the dead. But it made me feel alive to know they were there.

18

Club Aurora

It was minus six degrees when I woke up. Manterfield and I were sharing the tent, yet I never saw him arrive or depart, and I wondered if he'd stayed up all night. White frost covered the interior of the tent fabric and the rim of my sleeping bag. The tent door was frozen solid. My boots sat just inside, where Manterfield had moved them to keep them warm.

Outside, the dogs lay curled in little balls. The sky was deep blue and the sun shone down like a spotlight. Jason and Mayah had withstood an extreme team-building experience the night before and were giggling in their tent. The white world surrounding us was peaceful and quiet, with only a slight groan of sea ice to disturb the silence.

I climbed a low ridge behind the hut and saw Manterfield and Scott walking down. They had been up since four thirty, hiking the mountain to watch the sunrise. I found a slab of granite to sit on with a clear view of Erik the Red's peak. A cloud bank directly east appeared to be moving our way. Snow squeaked under my feet. Shadows from the mountains were not gray; they were pale, like the blue pools of seawater gathering on the ice.

Back at the hut, Manterfield, Scott, Mugu, Justus, and Mikael were drinking coffee. It was a surprisingly civilized scene, with a

At night, the hunters tethered their dogs roughly in an arc, creating a barrier between potentially curious polar bears and our sleeping quarters.

topped-off French press and conversation about Thomas Pynchon. Beneath Scott's ever-deepening layers was a literary scholar, Dartmouth grad, and major Pynchon fan. He had been a member of a Pynchon email group for thirty years. He shared his thoughts about the author's eschewal of the postwar establishment and argued how a postmodern master could call himself a classicist. I listened, in awe of this red-faced libertarian with the chiseled cheekbones and broad shoulders of New England, Anglo-Saxon stock. An example of his robust mind was a twenty-minute monologue, after the Pynchon critique, about how he whittled the number of keystrokes it takes to properly edit a time-lapse in Adobe Photoshop from sixty-five to twelve.

On the other side of the shack, Manterfield continued to search for his identity. He was locked in a power struggle with the hunters.

They thought they knew best about everything—they probably did—yet Manterfield knew best about what his guests needed. He definitely did. Justus, Mugu, and Mikael made a point of not listening to any command or request he made, smiling coyly while ignoring him. When he tried to help with the dogs, they brushed him off. When the pot for melting snow on the stove ran dry, they barked at him to fill it up. This was followed by Justus tirelessly, mirthfully yelling Manterfield's nickname at his dog all day.

Manterfield took charge after breakfast, ordering us to take the flies off our tents to start drying them. It was pure mountain guide protocol—foresee a complication before it happens—even if our expedition had no objective, no itinerary, and no real danger other than running out of our preferred freeze-dried dinner. Karma comes for us all, and after roasting Manterfield all morning, Justus flipped his sled in a stunningly violent collision with an ice bulge when he was driving out of camp. Our gear flew through the air. Justus flew through the air. Manterfield had, thankfully, suggested that we hop off for the tricky navigation, and I, also thankfully, had filmed the entire thing. (Manterfield's mistreatment at the hands of the drivers was blatant enough that I sent him the video file as a parting gift, collateral for future roasting.)

Justus nursed a smashed forearm for a few minutes, then loaded us up and we took off again. We were near the turnaround point of the trip but had made such good time that Manterfield was considering going further, into *tunu*. Our course was due north along the coast of Tiniteqilaaq into Ilertaa Sound. The ice was windswept again and fast, and we made the next hut on the southeast coast of Torteeinat in a couple of hours. We would sleep indoors that night, a comfort to all and a bonus that boosted our confidence to push still farther north toward Sermiligaaq and *tunu,* where no dogsled group from Pirhuk had ever ventured. The landscape was different on the other side of the pass, somehow more remote. There were no signs of life anywhere—no wind or color or sound of any kind. Sunshine from above and below felt warm on my cheeks. A long slope led to a frozen inlet packed with icebergs. The mountains ahead looked like the Swiss Alps. They were, in fact, named the Schweizerland Alps, Manterfield said, christened by a Swiss geophysicist-explorer in 1912.

The peaks blocked the wind on the long, white expanse leading up to them. Sweat rolled down Justus's neck as he concentrated on the lark's foot and a giant knot in the trace lines. "Don't like heat," he said. "Cold coming." He was right, the cold did come as the sun fell behind the peaks and a blast of freezing air rushed down the mountains. It arrived like a rush of water, thick and heavy, settling on the inlet, changing the air and the light in a matter of minutes. The temperature dropped twenty degrees. Mayah froze instantly and courageously toughed it out on a half-mile walk with Manterfield to warm up.

The going was slow all the way back to the hut. I hopped off on the last hill and hiked to get my blood flowing. It helped a little, but the chill was different from what we had experienced so far. There were no clouds to hold the heat in. My jacket didn't seem to be retaining my body heat, either.

Justus finally crested the pass and flew down the other side. We pulled into camp first, and he staked out the dogs as the other teams and frozen passengers arrived. Knowing the only way to stay warm was to keep moving, I grabbed a shovel and dug the latrine for an hour, then dug a built-in snow-couch with armrests and drink holders to watch the northern lights. Mayah and Jason joined me there after we finished dinner. For a couple constantly in awe, the light show and snowy sectional were almost too much. We played lounge music on a phone, illuminated the pit with a red headlamp, and christened it Club Aurora as the braided green rope wove itself across the sky. For the next hour it whipped the horizon, knitted itself into itself, and morphed into a symphony of phantasmic, pulsating clouds.

A blinking white light on the horizon interrupted the show. It blinked a few times, then disappeared, then appeared again way out on the sound. We hadn't seen another soul, building, or human-made object for days and were jarred by the thought of someone stumbling across us. As we informed Manterfield back in the hut, there was a knock at the door. I opened it and saw a man with ski goggles and a Pepsi in his hand, flanked by two other massive men and two snowmobiles. "Are you cold?!" he asked. All three men burst into laughter, and Justus jumped out of bed to greet them.

The men talked loudly as we went to bed. They had covered what took us three days in forty-five minutes. They were gone in the morning. Half of the group was still asleep; even Manterfield was snoring. I got up slowly and pulled on my down pants and jacket, then climbed the pass we'd sledded over the day before. I followed Mikael's tracks on the way up—his stylish moon boots left an obvious, perfectly circular print—and sat on a flat rock overlooking the inlet and the cutout peaks of the Schweizerland Alps.

This was the last stop on the expedition and the last one on my quest. It seemed as if we could go no farther north, though a hundred years before, Rasmussen, Nansen, Peary, and others had proven that we definitely could. This was the northernmost civilization on the planet, the last band of life in the Northern Hemisphere. It was also the fulcrum of one of the last tipping points on our warming planet.

A buzzing sound jolted me from my meditation. It took a minute to figure out where it was coming from. It was my phone; I hadn't looked at it in days. Messages were pouring in, and I realized that Mikael's tracks led to this ridge so he could use a cell signal to chat with his girlfriend. The messages on my phone were not cute. They came in a flurry of exclamation points and all caps. Several people had been trying to get in touch. The first message read: GET HOME NOW.

The First Wave

An old high school friend had sent the message. He had been on the first India trip I took for *Powder,* and we had climbed and skied all over the world. When he settled down and had a family, he became my default logistics contact for expeditions. It took a few texts before I understood the situation. The coronavirus was spreading at a new pace, and the United States had quickly become its breeding ground. That morning, the president had banned all travel into the United States from the European Union. I was not in the European Union, but my connection in Iceland definitely was. If I did not get home by tomorrow, I would likely be stuck in Greenland until the pandemic was under control.

"There is an expression in Spanish: 'God always forgives, we forgive sometimes, but nature never forgives'... I don't know if these are the revenge of nature, but they are certainly nature's responses." —*Pope Francis*

The prospect of staying on didn't sound that bad at first. Hunkered down in a self-sufficient, fully stocked base camp in the Arctic Circle with a former British commando seemed a good way to weather the storm. But the thought of Sara and Grey stuck in the middle of the coronavirus epicenter of the world made me sick to my stomach. I stashed my phone and followed Mikael's footsteps to the shed and another steamy, coffee-fueled breakfast session. It took a few minutes to find an opening in the conversational crossfire, and more than once I thought about not telling anyone. But we all had families and jobs, and it seemed the right thing to do.

You'd be hard pressed to find a better conversation stopper than telling a crowd gathered thousands of miles from their homes that the US president just closed their country's borders and that they had twenty-four hours to get back. Manterfield, seeming annoyed that I had information that he did not, immediately called Pirhuk HQ on the satellite phone. The hunters went back to sleep as each couple huddled, strategizing their plan of action. It was a scene worthy of a *Survivor* finale, complete with whispering, white knuckles, forced smiles, overly calm declarations, and lots of definitive nodding.

Eventually, everyone stated their intentions in a happy-go-lucky, I-don't-really-care-but-this-is-definitely-the-way-to-go tone, and a plan was hatched. We would take advantage of the fast ice and go straight to Kulusuk that day, stopping at the seal net on the way to justify the "hunt." It had taken us three days to get to this far north. On hard ice, Justus said, it would take two hours to get home.

There was indeed a tragically cute seal caught in the net—which I was told would be my charge for the journey home. Justus set it on top of a duffel bag and motioned for me to sit on it and not let it fall off. I grabbed the thing's frozen snout as we took off, trying to keep it, and me, from torpedoing off the sled. The expedition played before our eyes in high-speed reverse that afternoon. The maze of inlets, Star Wars icebergs, red hut, yellow hut, low passes, and the big descent that we now had to climb up. Justus was ruthless with his dogs. He could smell the warmth and comfort of his home and imagine his cute kids sitting on his lap. Coming down the last slope onto Torsuut Tunoq, he stayed on the front of the sled—no brakes!—skittering twenty miles an hour downhill. We went so fast we started to overtake the rear dogs in the pack, two of which tripped and barely missed being run over. Just as it looked like we were going to run over the entire team, the slope leveled out.

Manterfield didn't speak the entire ride. He seemed frustrated that the trip was cut a day short. He stared ahead as we rounded a long, low ridge and the airport came into view. There was the Kulu-

The Seal of Disapproval. This was my charge for the trip home. Every bump, every turn, I had to hold its cute, frozen snout and tail to keep both of us from flying off the sled. Two hours up and down mountain passes, over fjords and headlands. It was a miserable chore, yet by the end of the journey a kind of macabre intimacy had grown between us.

suk Hotel. There was the path leading to town. There was the stubbled outline of the village. We pulled in at sundown, and a few of the hunters' children came out to meet us. It took Justus about a minute to get our stuff off his sled and take off toward his house. We moved a bit slower, dragging our gear up the steps to the lodge, cracking the door, and practically falling into the comfort of the gear room. It was warm and dry. Everything looked soft. I could hear Theresa in the kitchen, making an unexpected dinner for ten. I hadn't taken off my base layers or fleece pants in four days. I lay them on the drying racks and grabbed a freshly laundered towel. As I waited for the water to warm up, I heard the rumble of what was rumored to be the last flight out of Greenland.

19

This World Is Brutal—Be Happy You Are Not Dead Yet

The flight was not the last out of Greenland. Or maybe it was. It was impossible for Helen to tell with the mixed messages coming from the airline, the airport staff, and even the government of Greenland, which appeared to want all foreigners off the island as soon as possible. The coronavirus was out of control. Millions were infected in China, Korea, Italy, the suburbs of Seattle, and New York City. Entire neighborhoods in Iceland had been quarantined.

The reaction worldwide was surprising, mostly in how unprepared citizens and governments were for a pandemic that doctors, authors, and hundreds of epidemiologists had warned about for half a century. There were not enough hospital beds, ventilators, nurses, or even surgical masks. The governor of New York set up army tents and cots in Central Park to handle the surge. A man in New Jersey hoarded a warehouse full of eighty thousand masks, offering to sell them at an absurd markup. (The mask scheme was quite believable for anyone from the New York area.) In this modern, digitized, internet-connected world, it was shocking to see entire nations floundering, reversing their messaging, resorting to draconian measures like allowing older patients to die so younger ones could be treated—a form of senicide similar to what Inuits once practiced.

As I scrolled through the news app on my phone for the first time in a week, it became clear that the world was ending. All the pomp, technology, military prowess, social awareness, and general progress of the modern world was a ruse. We were sitting ducks—cave people in the Appenzell Alps, peasants in medieval Venice, ice skaters of the 1600s on the frozen Thames—about to be unceremoniously wiped out by a plague.

Another analogy a few brave pundits began making pointed to a different planet killer that experts had been warning about for a half century while we watched. Climate change didn't have to be a crisis. If you accept the science and take action, you will find a much safer, more comfortable way. But the human brain did not evolve to register such awareness. We are incapable of calculating distant danger; our cognitive system responds only to risks right in front of us. Like a house on fire. Or a flood. John the Apostle put it this way (John 4:48): "Then said Jesus unto him, Except you see signs and wonders, you will not believe."

I had a hard time believing anyone that afternoon as the lodge descended into hushed chaos. At one point we were all going to pack up and make the new "last" flight out. An hour later that plane had departed and the new plan was to stay in Greenland indefinitely (again, not a terrible choice considering the multiple box freezers full of food in the basement). By dinnertime, half of the group's flights to the United States had been canceled, and we scrambled to book new ones through the United Kingdom, which because of its indeterminate situation with Brexit was somehow still allowed to fly to the United States. The travel ban had gone into effect, but US citizens were now allowed to return home to a half dozen airports, where they would be screened. Anyone with a fever would be detained in a "federal quarantine." Spenceley's suggestion: pack Tylenol so you don't get abducted by the CDC (Centers for Disease Control and Prevention).

Helen took on the role of dispatcher, seated at the kitchen counter, phone in hand, laptop whizzing through airline websites and news channels. Keystrokes got faster, tapping became hammering. Helen's very British way of grinning and bearing whatever troubling news she had just absorbed began to crack. The kitchen door closed. The entire Pirhuk team huddled and whispered. Big decisions were being made. Something we did not know about was on the line. It was not just our trip that was affected. We had been the first guests of the season; there were dozens of others either on their way or sitting at home deciding if they should cancel.

Manterfield processed the news with profound sadness—head hung, feet shuffling through the tight hallways. He did not get worked up; I never saw him on the phone. He went about his daily chores with the same easygoing attitude, maybe annoyingly so to his colleagues as he drifted through the kitchen and spoke softly to guests in the dining room about what may or may not happen, who knows, we'll be all right—*how about some more tea?* Spenceley bore the weight of the crisis. Should they cancel the season? Would guests cancel anyway? How would he pay off the new lodge? One Argentinian man on his way to Pirhuk was stranded in Spain. His connecting flight to Greenland had been canceled. Then Argentina closed its borders, trapping him in a hotel room with a mandated quarantine. Another group was stuck in Iceland, unsure if they should wait for a chance flight to Greenland or return home before their countries shut their borders.

It was a tricky line to toe, the same that Spenceley had been flirting with since he brought his first guests to the island. Greenland was probably better off without tourists, yet the jobs, fees, and income Pirhuk offered the community helped keep it afloat. Spenceley pleaded his case to various officials, as Helen searched for flights for the four hundredth time. The bureaucrats were not interested; borders were closed, and Pirhuk's season was over.

What happened next was a bit of a blur. Theresa vanished without a word the next morning. Having found a flight for herself, she was off to the United Kingdom to be with her family. Scott and Wendy had spent most of the previous evening rebooking flights, then took off early with Manterfield, Justus, Mayah, and Jason to make up for the lost day of dogsledding. It was impressive, even a bit odd, I thought, considering the situation. I receded into doomscrolling in a corner of my room, where I got a single bar of cell phone reception. Sara was scared and uncharacteristically paranoid. The virus could be contracted from a lamppost, from takeout packaging, from Grey's beloved swings on the playground. The virus was airborne; hundreds of people a day in New York City were dying from it. Sara hadn't left the apartment in a week. She wasn't sure how to get food. She didn't think I would be allowed back into the country, and even if I was, I would probably catch the virus on the plane and bring it home. I promised her that I would make it, that we would go to the cabin, hide out for a few weeks, and return to the city when things calmed down.

This vision became my mission. I visualized it as I packed my bags for the third time, as we ate our final dinner at Pirhuk, as I showered the final morning and tucked two surgical masks my mother had given me weeks before into my pocket, along with a bottle of hand sanitizer and four Tylenol capsules. Kulusuk Airport was packed: ours would in fact be the last flight out of Greenland for the foreseeable future. Every seat was taken. The woman at the french fry machine was manic — too many people, too many orders. An entire grade school class marched into the lobby and began yelling and running around. Old women sneezed; a nervous middle-aged man leaned against the bathroom wall sniffling into a handkerchief. When the security gate opened, I eased into the departure lounge and spent an hour staring at seal pelts.

The last human I saw before boarding the flight was Peter, the snowmobile driver who had brought us a Pepsi on our last night in the hut. He worked at the airport and took my ticket at the door, offering a grin that distinctly said *this world is brutal—be happy you are not dead yet*. The pilot agreed. "Welcome to the last flight out of Greenland!" he chirped. Then he turned the cabin temperature down to freezing. I watched the snowy peaks around Kulusuk disappear and then the great glaciers that Rasmussen had explored and finally the ragged, green edge of the ocean, where Nansen and his men had navigated. Why did they do it? What drew them here? What drew me here? "I can tell you deliverance will not come from the rushing noisy centers of civilization," Nansen once said. "It will come from the lonely places."

After a layover in Iceland, I survived the CDC screeners at JFK and the crowds at customs and at baggage claim. At home, I wiped down my bags and hands with sanitary wipes and saw the most incredible sight: Grey and Sara sitting in the living room, waiting for me to arrive. This was the human world. The cryosphere was the ancient world. I walked in, washed and dried my hands and face, then scooped up my girls and settled on the couch for what would become a three-day lounge. It didn't seem real. The world was closing down; the ice sheets were melting; the coronavirus would change us forever. My trip to visit Kim Maltais again at the Sheep Ranch in Washington was off. I wouldn't be visiting Bird in Twisp or Chamonix, either.

Wailing ambulances raced up our street for the next few days. Sara and I finally packed what we could fit into the car and drove north. We moved into the cabin for a month, then two, then three. We split wood for heat, carried in our water. We took Grey out every day for walks in the woods. Weather whooshed down Overlook Mountain in waves—rain, snow, wind, flood, sometimes all at once with no warning. We listened to rain fall, then watched the stream swell above its banks as the sun came out. While we hid out,

View from the last flight out of Greenland. This was the new world, the media said repeatedly. The old world was gone forever. It was like the weeks after 9/11, the years after Watergate and the Vietnam War, and every other catastrophe—like climate change—that humans had experienced or invented. "The earth will survive," Koni Steffen had told me. "It's not a question. It has survived much bigger cycles, but it's more about human beings. Extinction is a hard word, but we have seen it, too. A big impact killed off 95 percent of all the life on earth. These histories are way, way beyond our imagination. What keeps me up at night more recently is the permafrost on Greenland. How long are we safe to do research?"

everything around us changed. New York City reopened, then shut down again. Koni Steffen died. David Reddick called to say that after forty-nine years, *Powder* magazine was being shuttered. Kelly Gleason, Jon Riedel, and most of the scientists I had spoken with were homebound—research trips scrapped, drilling operations stopped, classes canceled, climate change research put on hold. (Except for the indefatigable Seth Campbell, who, just before this book was completed, was handpicked by NASA to help engineer an ice-core drilling plan for the massive glaciers and ice aprons of Mars, on the first manned mission to the red planet, slated for the 2030s.)

One season led to the next: spring, summer, fall. Then the first snow of winter spindled down from the sky. We lived in the mountains now. We gave up our Brooklyn apartment. Grey went to a new school. In February, we headed to a local ski hill a few minutes away. In the shadow of the base lodge, I clicked her tiny boots into a pair of skis that Reddick had sent me when she was born. We skied all morning, her in a snowplow, me holding her by her armpits. We sat by a giant fireplace at lunch and ate pizza and french fries.

This was our new world, brought on by something we couldn't foresee or change. It was not as hard to adapt to as I had thought. Certainly not as hard as the end of winter will be. We had a new president, who rejoined the Paris Agreement, shut down oil pipelines, poured millions into renewable energy, and plotted a new course toward a more rational, hopefully cooler, future. Still, climate change has momentum, and I knew that winters would continue to shrink. The rain line would climb higher. Spring would start a bit earlier. Maybe decisive climate change legislation would save us from the worst-case scenario. Maybe the message that M3 interpreted from the Juneau Icefield would make its way to Washington, D.C. Maybe the thermohaline current would deliver a millennium of freezing winters and deep snow, and maybe Kim Maltais would someday see the connection between fire and snow.

It is impossible to say when the last winter will come. So Sara, Grey, and I enjoyed it while we could. The cold bite of a January morning. Snowy footprints in the living room. Skis stacked outside the door, and a few winter nights when crystal-filled clouds drifted through the dark sky and dropped a veil of snow that obscured the driveway, the woodpile, the muddy pathways, and all the imperfections of our human lives.

Acknowledgments

The creation of this book was a story in itself as the coronavirus pandemic chased me around the globe. I was sitting on an unnamed pass in the Arctic Circle when the first travel ban went into effect. After somehow getting home and absconding with my family to Upstate New York, I finished writing the first two parts of the book, "The Fires" and "The Icefield," sequestered in a sailboat—first on land and then, thankfully, in the water. Part 3, "The Alps," coalesced in our guest cabin, then in a car parked in the driveway of that cabin. Part 4, "White Earth," was completed in a converted shipping-container office, aptly, in the freezing dark days of winter—with a car battery to run my monitors and a cord of wood to keep me warm. My first thanks go to all who helped provide these spaces in the most difficult year.

A generous grant from the David Rockefeller Fund made the reporting for this book possible, and I am deeply indebted to Lukas Haynes for recognizing the importance of snow, ice, and winter to our world and our lives. I also thank Lisa Pohlmann and the Natural Resources Council of Maine for their support as well as Seth Campbell, Michael Zemp, Anne Nolin, Auden Schendler, Steve Tatigian, Ned Hutchinson, Kade Krichko, Tophie Kane, Jamie Ziobro, the JIRP crew, Agustina Lagos Marmol, Dr. Grace Gronkowski, and the Pirhuk crew for providing contacts and early logistical support that shaped the initial web of characters and storylines.

The initial vision of my heroic agent, Duvall Osteen, and editor, Josh Kendall, not only got this book rolling but also nudged it toward its final form while the world appeared to fall to pieces. I have found this kind of sustained support, focus, and perseverance extremely rare in the publishing world and am forever in debt and in awe of their sincere, insightful, and dedicated work. Finally, I thank David Reddick, Kevin Back, Matt Hansen, the Moe brothers, and all the staff at *Powder* magazine for keeping the ski-bum flame alight and feeding my fantasy of exploring snowy climes and writing about them for two decades. Your legacy will never be forgotten. *Powder to the People!*

As always, the real stars of the show are my wife, Sara, and daughter, Grey, who lived with a maniac for eighteen months and kept our ship afloat through the disaster that was 2020. You ladies are the best.

Suggested Readings

Most of the scientific papers and data referenced in this book can be easily found online, and I urge readers to delve into them to understand (1) the incredible work being undertaken by scientists today and (2) the raw data they have uncovered. Armed with this information, you can make your own inferences about where we and our planet are headed. The Intergovernmental Panel on Climate Change and other UN climate and cryosphere reports are a good place to start. Publications from the World Glacier Monitoring Service are another excellent resource, including the service's phone app, on which you can see the real-time health of every major glacier in the world.

When talking about the forests of the Pacific Northwest, I instinctively point readers to John Vaillant, *The Golden Spruce: A True Story of Myth, Madness, and Greed* (W. W. Norton, 2006), for its unmatched storytelling and reporting on the creation and destruction of the planet's largest concentration of biomass. Gary Ferguson, *Land on Fire: The New Reality of Wildfire in the West* (Timber Press, 2017), offers another excellent primer on the expanding fire season in the US West. For a more impressionistic view of the Cascades, check out John Muir, *Steep Trails: Explorations of Washington, Oregon, Nevada, and Utah in the Rockies and Pacific Northwest Cascades* (Houghton Mifflin Company, 1918).

John Muir, *Travels in Alaska* (Houghton Mifflin, 1915), offers excellent early observations of the glaciers around Juneau, Alaska—observations that you can compare with the writings of Maynard

Malcolm Miller. (Both authors' work can be found online.) Klaus Dodds, *Ice* (Reaktion Books, 2018), and Bernd Brunner, *Winterlust: Finding Beauty in the Fiercest Season* (Greystone Books, 2019), provide lyrical and intriguing big-picture histories of snow, ice, and winter culture, while publications and reports from the National Snow and Ice Data Center in Boulder, Colorado—along with materials from NOAA and NASA—are terrific online resources for anyone interested in the formation, health, and current status of the cryosphere.

The study of the Alps is the study of human migration upward. There is no better source, format, style, or even concept of how to cover this gargantuan history than the works of Jon Mathieu. Start with *The Alps: An Environmental History* (Polity, 2019) for a complete archive of the range unlike any you have ever read. From there, move on to *The Third Dimension: A Comparative History of Mountains in the Modern Era* (White Horse Press, 2013) to get an unmatched glimpse of the high peaks and their ecology and heritage. The only comparable work, in terms of transcendence, is Maria Savi-Lopez, *Leggende Delle Alpi* (*Legends of the Alps*) (Torio E. Loescher, 1889), a meticulous history of demons and dragons in the Alps. Susan Barton, *Healthy Living in the Alps: The Origins of Winter Tourism in Switzerland, 1860–1914* (Manchester University Press, 2014), retells another bizarre storyline—of the transition from medical tourism in the Alps in the 1800s to the winter tourism that defines the range today.

Greenland could have its own library here. Anyone interested in the great island should start with the literary tradition of its first explorers. Fridtjof Nansen, *The First Crossing of Greenland* (Longmans, Green, and Company, 1890; paperback ed. University Press of the Pacific, 2001), is a page-turner, as are Knud Rasmussen's many books about his travels in the Arctic Circle, including *Across Arctic America: Narrative of the Fifth Thule Expedition* (G. P. Putnam's Sons, 1927; paperback ed. University of Alaska Press, 1999). Peter Freuchen's books, such as *Arctic Adventure: My Life in the Frozen North* (Farrar & Rinehart, 1935; paperback ed. Schuyler Press, 2011), add a touch of horror to the Thule expedition tales, as he recounts some of the deadlier moments of the journey. For a more modern, though equally heroic, account of travel through Greenland, be sure to read Gretel Ehrlich's astonishing and insightful *This Cold Heaven: Seven Seasons in Greenland* (Pantheon, 2001).

Index

Note: Italic page numbers refer to illustrations.

About the Author

Porter Fox was born in New York and raised on the coast of Maine. He is the author of *Northland: A 4,000-Mile Journey Along America's Forgotten Border* and *Deep: The Story of Skiing and the Future of Snow*.

Fox lives, writes, and teaches in Upstate New York, where he also edits the award-winning literary travel writing journal *Nowhere*. His work has been published in the *New York Times Magazine*, the *Believer*, *Outside*, the *Virginia Quarterly Review*, *Men's Journal*, and *Powder*, and has been anthologised in *The Best American Travel Writing*. He won a Western Press Association award in 2014 for a two-part feature about climate change and a Lowell Thomas Award for an excerpt from *Northland*.

Fox is a MacDowell Colony fellow and teaches at Columbia University School of the Arts.